THE SPSS GUIDE TO THE NEW STATISTICAL ANALYSIS OF DATA

Springer
*New York
Berlin
Heidelberg
Barcelona
Budapest
Hong Kong
London
Milan
Paris
Santa Clara
Singapore
Tokyo*

THE SPSS GUIDE TO THE NEW STATISTICAL ANALYSIS OF DATA

by T.W. Anderson and Jeremy D. Finn

Susan B. Gerber
Kristin E. Voelkl

State University of New York at Buffalo

Springer

Susan B. Gerber
State University of New York at Buffalo
Graduate School of Education
Buffalo, NY 14260-1000
USA

Kristin E. Voelkl
State University of New York at Buffalo
Graduate School of Education
Buffalo, NY 14260-1000
USA

Library of Congress Cataloging-in-Publication Data
Gerber, Susan B.
  The SPSS guide to The new statistical analysis of data by T.W.
Anderson and Jeremy D. Finn/Susan B. Gerber, Kristin E. Voelkl.
    p.   cm.
  Includes bibliographical references (p.   ).
  ISBN 0-387-94821-X (softcover:alk. paper)
  1. Statistics—Data processing:Programmed instruction.   2. SPSS
for Windows—Programmed instruction.   I. Anderson, T.W. (Theodore
Wilbur), 1918–. The new statistical analysis of data.   II. Voelkl,
Kristin E.   III. Title.
QA276.4.G46   1997
519.5′078′55369—dc21   96-47394

Printed on acid-free paper.

© 1997 Springer-Verlag New York, Inc.
All rights reserved. This work may not be translated or copied in whole or in part without the written permission of the publisher (Springer-Verlag New York, Inc., 175 Fifth Avenue, New York, NY 10010, USA), except for brief excerpts in connection with reviews or scholarly analysis. Use in connection with any form of information storage and retrieval, electronic adaptation, computer software, or by similar or dissimilar methodology now known or hereafter developed is forbidden.
The use of general descriptive names, trade names, trademarks, etc., in this publication, even if the former are not especially identified, is not to be taken as a sign that such names, as understood by the Trade Marks and Merchandise Marks Act, may accordingly be used freely by anyone.

Production managed by Hal Henglein; manufacturing supervised by Joe Quatela.
Camera-ready copy supplied by the authors.
Printed and bound by Braun-Brumfield, Inc., Ann Arbor, MI.
Printed in the United States of America.

9 8 7 6 5 4 3 2 1

ISBN 0-387-94821-X Springer-Verlag New York Berlin Heidelberg   SPIN 10541587

Preface

This book is a self-teaching guide to the SPSS for Windows computer package. It is designed to be used hand-in-hand with *The New Statistical Analysis of Data* by T. W. Anderson and Jeremy D. Finn, although it may be used as a stand-alone manual as well. This guide is very easy to follow since all procedures are outlined in a straightforward, step-by-step format. Because of its self-instructional nature, the beginning student can learn to analyze statistical data with SPSS without outside assistance. The reader is "walked through" numerous examples that illustrate how to use the SPSS package. The results produced by SPSS are shown and discussed in each application. The data sets used in the examples are primarily those in the Anderson-Finn textbook.

To the extent that SPSS performs the procedures in the Anderson-Finn text, this manual follows the book chapter-by-chapter. Except for Chapter 1, the section titles in the manual correspond to sections in the textbook. Each chapter demonstrates the statistical procedures described in the textbook and gives exercises that can be performed for further practice. (An instructor may choose to use these as additional class assignments.)

This manual was created using SPSS for Windows, Version 6.1, on an IBM-compatible personal computer. There are other versions available, including: Version 7.0 for Windows 95, Version 6.1 for Macintosh, and Version 6.1 for Power Macintosh. There are small differences among these versions; for example, the appearance of the screens is slightly different in the Macintosh versions. However, SPSS Corporation indicates that the syntax and general procedures are virtually identical. Therefore, Macintosh and Windows 95 users will also be able to use this manual.

SPSS for Windows is readily available in two forms: the Graduate Pack and the Student Version. The Graduate Pack is the complete program; with it, the user can perform all analyses available on SPSS. The Student Version is limited in the number of variables (50) and cases (1500) it can accommodate. In addition, with the Student Version the user cannot perform all advanced statistical procedures available with the Graduate Pack. All the procedures illustrated in this manual are available with both programs, however. Because complete manuals are not included with either package, this guide may prove to be particularly useful.

In order to get the maximum benefit from this manual, the student should run the examples using SPSS as he/she reads through the guide. This will require a computer with Microsoft Windows version 3.1 or later and one of the versions of SPSS listed above installed before beginning. Information about installing the program is provided with the software when it is obtained. The user should be familiar with using the mouse, although it is also possible to use the keyboard for any procedure. Chapter 1 of this guide describes how to start the SPSS program and how to open a data file. In Chapters 2 through 16 are the descriptions of statistical procedures which assume that a data file has been opened.

. An appendix has been added to each chapter which provides syntax to be used in running statistical procedures from the SPSS Syntax Window. SPSS syntax is the command format used in non-window versions of SPSS and those running on mainframe computers. While this manual is intended to be used with the pull-down menus, readers familiar with SPSS syntax may wish to use this command format with SPSS for Windows as well. The only syntax we have provided is the SPSS command required to run a procedure. These commands assume that the user has opened an SPSS data file before executing the commands. For further information about SPSS syntax, the reader may consult the *SPSS for Windows Base System User's Guide*.

The data files used in this manual are posted on the Springer-Verlag website on the Internet. All 36 data sets are available for viewing and can be easily downloaded from the Internet to the user's local system. The files can be found at the following Internet address: http://www.springer-ny.com/supplements/gerber. Users lacking web access can use FTP to retrieve the data sets by typing "ftp://ftp.springer-ny.com/pub/supplements/gerber." Individuals who do not have access to the World Wide Web can obtain a copy of the data sets on diskette by writing to Springer-Verlag customer service at custserv@springer-ny.com or by phoning 1-800-SPRINGER.

Instructors may wish to download the data sets to a public site on their local computer server so that students can simply use the data on the local network without having to access the Internet. When downloading to the hard drive, it is suggested that the user choose a suitably named directory in which to save the data sets. For example, to download a data file to the c:\data\ directory: (1) Use your web browser to locate the Springer URL (http://www.springer-ny.com/supplements/gerber). This will place you at the Springer-Verlag website. It should read "The SPSS Guide to the New Statistical Analysis of Data" at the top of the page. (2) Scroll down and click on the link that reads "Download Data Sets." (3) Point your mouse arrow to the data file that you wish to save. To download the file, "right-click" on the file. (Notice that if you "left-click" the data file will open for you to view on the screen.) Click on the option "Save this Link as..." to open the "Save As" dialog box. (4) In the Drive and Directory windows, choose the appropriate location where you wish to save the data. In this example, the Drive is "c:" and the Directory is "data." (5) Click on OK. Repeat this process for all data files you wish to download. The data files will be saved at the specified location on your local hard drive or server and are now available to be read by SPSS as illustrated in Chapter 1 of this guide.

Each data set is provided in two forms, as a raw data ASCII file, and as a file produced by SPSS. While both files have the same file name, the ASCII file has the extension ".dat" and the SPSS files have the extension ".sav" (for example, "final.dat" and "final.sav"). A description of each ASCII file is given in the Appendix. This includes column locations and formats for each variable in the ASCII file and the names of the variables used in the examples and contained on the SPSS data files.

The reader will find it easiest to use the SPSS ".sav" data files in running the examples described in this manual. The ASCII files may be used as well; indeed, many statistical data sets may be prepared first in the form of an ASCII file. When an analysis begins with an ASCII file, the researcher needs to describe the characteristics of the data set to SPSS; instructions for doing this are given in Section 1.2.1 of this manual.

SPSS for Windows includes a useful Help facility; the user only needs to click on **Help** on the main menu bar. For more detailed information about SPSS for Windows, the reader may consult the *SPSS for Windows Base System User's Guide*, or call SPSS technical support at (312) 329-3410. Finally, the e-mail address for technical support is support@spss.com.

The authors have invested a great deal of energy to produce a manual that is informative and easy to understand. If readers have any suggestions for future editions, we would greatly appreciate receiving them.

S.B.G.
K.E.V.
August 1996

Contents

Part I: Introduction — 1

Chapter 1 The Nature of SPSS — 3

1.1 Getting Started with SPSS for Windows — 3
1.2 Managing Data and Files — 7
1.3 Transforming Variables and Data Files — 15
1.4 Examining and Printing Output — 23
1.5 Missing Values — 24
Chapter Exercises — 25

Part II: Descriptive Statistics — 27

Chapter 2 Organization of Data — 29

2.2 Organization of Categorical Data — 29
2.3 Organization of Numerical Data — 33
Chapter Exercises — 39
Appendix II: SPSS Syntax for Organization of Data — 40

Chapter 3 Measures of Location — 41

3.1 The Mode — 41
3.2 The Median and Other Percentiles — 43
3.3 The Mean — 44
Chapter Exercises — 48
Appendix III: SPSS Syntax for Measures of Location — 50

Chapter 4 Measures of Variability — 51

4.1 Ranges — 51
4.2 The Mean Deviation — 52
4.3 The Standard Deviation — 53
4.5 Some Uses of Location and Dispersion Measures Together — 54
Chapter Exercises — 56
Appendix IV: SPSS Syntax for Measures of Variability — 58

Chapter 5 Summarizing Multivariate Data: Association Between
          Numerical Variables   59

          5.1 Association of Two Numerical Variables   59
          5.2 More than Two Variables   67
          Chapter Exercises   68
          Appendix V: SPSS Syntax for Summarizing Multivariate Data:
              Association Between Numerical Variables   71

Chapter 6 Summarizing Multivariate Data: Association Between
          Categorical Variables   73

          6.1 Two-by-Two Frequency Tables   73
          6.2 Larger Two-Way Frequency Tables   77
          6.3 Three Categorical Variables   79
          6.4 Effects of a Third Variable   83
          Chapter Exercises   86
          Appendix VI: SPSS Syntax for Summarizing Multivariate Data:
              Association Between Categorical Variables   88

Part III: Probability   89

Chapter 7 Basic Ideas of Probability   91

          7.3 Probability in Terms of Equally Likely Cases   91
          7.8 Random Sampling; Random Numbers   93
          Chapter Exercises   93
          Appendix VII: SPSS Syntax for Basic Ideas of Probability   94

Chapter 8 Probability Distributions   95

          8.5 Family of Standard Normal Distributions   95
          Chapter Exercises   97
          Appendix VIII: SPSS Syntax for Probability Distributions   98

Chapter 9 Sampling Distributions   99

          9.1 Sampling from a Population   99
          9.2 Sampling Distribution of a Sum and of a Mean   100
          9.5 The Normal Distribution of Sample Means   101
          Chapter Exercises   103

Part IV: Statistical Inference — 105

Chapter 10 Using a Sample to Estimate Characteristics of One Population — 107

10.1 Estimation of a Mean by a Single Number — 107
10.2 Estimation of Variance and Standard Deviation — 109
10.3 An Interval of Plausible Values for a Mean — 109
10.4 Estimation of a Proportion — 112
10.5 Estimation of a Median — 112
10.6 Paired Measurements — 113
Chapter Exercises — 114
Appendix X: SPSS Syntax for Using a Sample to Estimate Characteristics of One Population — 115

Chapter 11 Answering Questions About Population Characteristics — 117

11.1 Testing a Hypothesis About a Mean — 117
11.3 Testing Hypotheses About a Mean when the Standard Deviation is Unknown — 119
11.4 P Values: Another Way to Report Tests of Significance — 121
11.5 Testing Hypotheses About a Proportion — 122
11.6 Testing Hypotheses About a Median: The Sign Test — 124
11.7 Paired Measurements — 126
Chapter Exercises — 129
Appendix XI: SPSS Syntax for Answering Questions About Population Characteristics — 131

Chapter 12 Differences Between Two Populations — 133

12.1 Comparison of Two Independent Sample Means when the Population Standard Deviations are Known — 133
12.2 Comparison of Two Independent Sample Means when the Population Standard Deviations are Unknown but Treated as Equal — 134
12.3 Comparison of Two Independent Sample Means when the Population Standard Deviations are Unknown and not Treated as Equal — 137
12.4 Comparison of Two Independent Sample Proportions — 137
12.5 The Sign Test for a Difference in Locations — 139
Chapter Exercises — 141
Appendix XII: SPSS Syntax for Differences Between Two Populations — 143

Chapter 13　Variability in One Population and in Two Populations　145

　　13.1　Variability in One Population　145
　　13.2　Variability in Two Populations　146
　　Chapter Exercises　148
　　Appendix XIII: SPSS Syntax for Variability in One Population
　　　and in Two Populations　149

Part V: Statistical Methods for Other Problems　151

Chapter 14　Inference on Categorical Data　153

　　14.1　Tests of Goodness of Fit　153
　　14.2　Chi-Square Tests of Independence　156
　　14.3　Measures of Association　158
　　Chapter Exercises　162
　　Appendix XIV: SPSS Syntax for Inference on Categorical Data　165

Chapter 15　Simple Regression Analysis　167

　　15.1　The Scatter Plot and Correlation Coefficient　167
　　15.2　SPSS for Simple Regression Analysis　169
　　15.3　Another Example: Inverse Association of x and y　175
　　Chapter Exercises　180
　　Appendix XV: SPSS Syntax for Simple Regression Analysis　181

Chapter 16　Comparisons of Several Populations　183

　　16.1　One-Way Analysis of Variance　183
　　16.2　Which Groups Differ from Which, and by How Much?　188
　　16.3　Analysis of Variance of Ranks　190
　　Chapter Exercises　192
　　Appendix XVI: SPSS Syntax for Comparisons of
　　　Several Populations　194

Appendix　Data Files　195

PART I

INTRODUCTION

Chapter 1 THE NATURE OF SPSS

1.1 Getting Started with SPSS for Windows

SPSS for Windows is a versatile computer package that will perform a wide variety of statistical procedures. To start SPSS for Windows, begin with the main windows screen and double click the SPSS icon in the Program Manager.

1.1.1 Windows

In running SPSS, you will encounter several types of windows. The window with which you are working at any given time is called the *active* window. The six types of windows are:

APPLICATION WINDOW. This is the main window. It is always open, but does not itself contain any data or output. It is like the control panel, consisting of the main menu bar and icon bar from which you open other windows and run statistical procedures.

DATA EDITOR WINDOW. This window shows the contents of the current data file. A blank data editor window automatically opens when you start SPSS for Windows.

OUTPUT WINDOW. This window displays the results of any statistical procedures you run, such as descriptive statistics or frequency distributions. An empty output window automatically opens when you start SPSS for Windows.

CHART CAROUSEL WINDOW. This window is similar to the output window but contains all the graphs and charts that you direct SPSS to create.

CHART WINDOW. If you wish to edit a chart, clicking on the edit button of the Chart Carousel Window will change it to the Chart Window. You will note that the menu and icon bars are different from those in the other Application Window. The options in this window are primarily used for editing the charts. You can, for instance, rotate axes, change the colors of charts, select different fonts, and the like.

SYNTAX WINDOW. You will use this window if you wish to write SPSS "syntax" to run commands instead of clicking on the pull-down menus. The advantage to this method is that it allows you to perform a series of commands by simply clicking on the Run button on the icon bar of the main menu. In computer terminology, this is anaologous to the difference beween working in batch mode (using syntax windows) and in interactive mode (using pull-down menus).

Figure 1.1 shows what the screen will look like when SPSS for Windows first opens. The full screen, with the menu entitled SPSS for Windows, is the Application window. Nested within that window is the Output window. The Data Editor window is the frontmost window.

Figure 1.1 SPSS for Windows Initial Screen

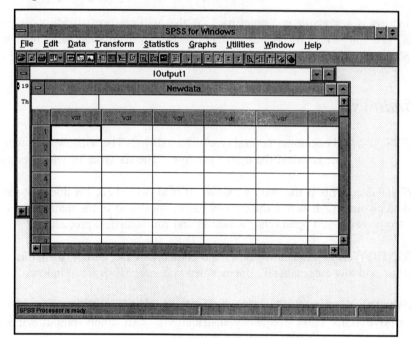

1.1.2 *Changing and Closing Windows*

The Application Window, a Data Editor Window, and an Output Window open automatically when you start SPSS. The window with which you are currently working is called the *active* window. There are three ways to change the active window:

(1) *Using Icons*. There are three icons on the icon bar that pertain to the different types of windows. They comprise the second group of icons, or the fifth, sixth, and seventh individual icons. The first of these allows you to cycle through output windows. That is, if you had several output windows open, clicking on this button would activate the output window next closest to the foreground of the screen. Similarly, the second of these icon buttons (sixth button overall) cycles through all open syntax windows, and the last button cycles through open chart windows.

(2) *Using Next*. You may click on the **minus sign** on the top left corner of the active window and select **Next** from the pull-down menu. This will move the current window to the last position in the series of open windows. You would continue this procedure, cycling through the open windows until the window you want is the active window. Note that with this option you cycle through all open windows, regardless of type, but with the icons option, you only activate windows of a given type.

(3) *Locating and Clicking*. When more than one window is open, the windows are displayed so as to overlap. That is, you are able to see the edges of windows that are open but not designated as active. You are able to change a window to active simply by clicking on the edge of the desired window.

Note that none of these options involves closing any windows. You are simply changing the order of the stack of open windows. In fact, you are not permitted to close the output or data editor windows; one remains open at all times. You may, however, close the syntax, chart, and chart carousel windows.

To close a window:

(1) Make the window the active window by one of the methods described above.

(2) Click on the **minus sign** on the top left corner of the window.

(3) Click on **Close** from the pull-down menu.

Despite the fact that you cannot close the actual Data or Output windows, you can change their contents. This has the same effect as clearing the window and "starting over." For example, you can open a new data file at any time during your working session.

To open a new data file:

(1) Click on **File** from the main menu.

(2) Click on **Open** from the pull-down menu.

(3) Click on **Data** from the secondary pull-down menu.

(4) Select the name of the data file you wish to open. (See Section 1.2.1 for details on how to open data files.) Before SPSS executes this command, you will be prompted as to whether you wish to save the contents of the current data file.

You may also create a new data file by following an alternative procedure. Instead of selecting **Open** from the pull-down menu (step 2), you would click on **New**.

You can also clear the contents of your output window using the same method:

(1) Click on **File** from the main menu.

(2) Select **New** from the pull-down menu.

(3) Select **Output** from the secondary pull-down menu. Again, you may be prompted as to whether or not you wish to save the contents of the current output window.

1.1.3 The Main Menu

SPSS for Windows is a menu-driven program. Most functions are performed by selecting an option from one of the menus. These menus are referred to as "pull down" menus since an entire menu of options appears when one is selected. The main menu bar is where most functions begin. It is located directly under the banner bar entitled "SPSS for Windows" (see Figure 1.1). Any menu may be activated by simply clicking on the desired menu, or using the Alt-letter keystroke (each menu uses the first letter in the menu word). For example, to activate the file menu, either click the mouse on **File** or use the keyboard with **Alt-F**. The main menu bar lists 9 menus:

FILE. This menu is used to create new SPSS files, open existing files, and read files that have been created by other software (e.g., spreadsheets or databases).

EDIT. This menu is used to modify or copy text from the output or syntax windows.

DATA. Use this menu to make temporary changes in SPSS data files, for example, merging files, transposing variables and cases, and creating subsets of cases for analyses. Changes are not permanent unless you explicitly save the changes.

TRANSFORM. The transform menu makes changes to selected variables in the data file and computes new variables based on values of existing variables. These transformations are not permanent unless you explicitly save the changes.

STATISTICS. Use this menu to select a statistical procedure to be performed such as correlations, analysis of variance, and crosstabulations.

GRAPHS. This menu is used to create bar charts, pie charts, histograms, and scatter plots. Some procedures also generate graphs.

UTILITIES. This menu is used to change fonts, display information on the contents of SPSS data files, or open an index of SPSS commands.

WINDOW. Use the window menu to arrange, select, and control the attributes of the SPSS windows.

HELP. This menu opens a Microsoft Help window containing information on how to use many SPSS features.

1.2 Managing Data and Files

Entering and selecting data files in SPSS for Windows is quite easy. We will first demonstrate how to open an existing data file, and then how to enter raw data from scratch.

1.2.1 Opening Data Files

SPSS for Windows can read different types of data files. The file type we will use in this manual are SPSS data files. These files are easily identified since (by default) each file name is followed by ".sav" extension. SPSS data files are unique in that they contain the actual data as well as information about the data such as variable names, formats, and column locations. These files are written in a special code that is read and interpreted by the SPSS program. If you try and read these data files with software other than SPSS, the file will look like lines of secret code and will not make sense to you. However, they make a great deal of sense to SPSS, and this is why reading them with SPSS is so easy. If you would like to look at the information contained in an SPSS data file (that is not currently open), you can do this by clicking on **File** in the menu bar, and then choose **Display Data Info**. Select the file name that you wish to examine and then click **OK**.

SPSS for Windows can also read raw data that are in simple text files in standard ASCII format. ASCII data files are usually identified by the ".dat" extension. These are data files that just contain ordinary numbers (or letters). There is no additional information contained in the file such as variable locations, formats, labels, missing values etc. (SPSS .sav data files do contain this additional information). You can read ASCII files with many different software programs including Microsoft Edit.

Note: Most of the examples in the following chapters use the SPSS data files that are provided on the Internet for use with this manual. Unless you are required to enter data on your own into a new file, all procedures assume that you have opened the data file before beginning any computations or analyses.

Reading SPSS Data Files

We will illustrate how to read an SPSS data file. The reader may follow along using the data on the Internet that accompany this guide.

To open a data file:

(1) Click on **File** in the menu bar.

(2) Click on **Open** on the file pull-down menu.

(3) Click on **Data** on the open pull-down menu. This opens the Open Data File dialog box as shown in Figure 1.2.

Figure 1.2 Open Data File Dialog Box

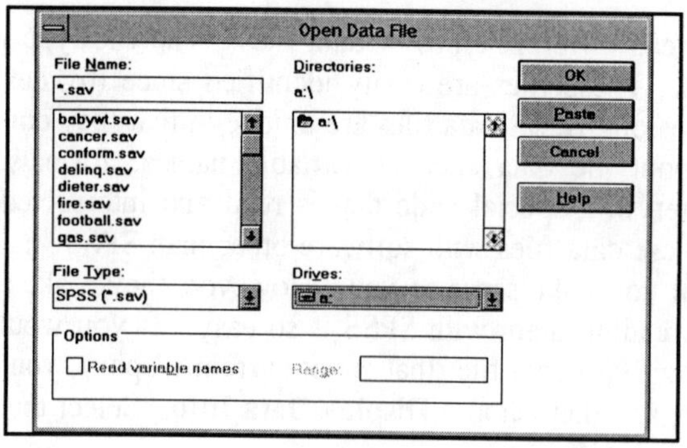

(4) Point the arrow to the data file you wish to open and click on it. By default, all SPSS data files (*.sav) in the current directory will be displayed in the list. If your data file is not visible in the file name box, use the up and down arrows to scroll through the files until locate your desired file. Note that all of the SPSS data files have the .sav extension, and this is designated in the File Type window. Before you open a data file, make certain that the designated file type, drive, and directory are correct. If you are reading SPSS data files and the file type box does not read "SPSS (*.sav)", you must scroll through the file types and select that type.

For example, to open the file called "football.sav," highlight the name of this file by clicking on it with the mouse button.

(5) Click on **OK** or press enter. You should now see the contents of the data file displayed in the Data Editor window. The "football.sav" data file contains two variables," height" and "weight," for 56 football players from Stanford University.

Reading ASCII Data Files

To read an ASCII data file, begin at the main menu bar:

(1) Click on **File**.

(2) Click on **Read ASCII Data**. This opens the Read ASCII Data File dialog box as shown in Figure 1.3.

(3) Select the file type "ASCII (*.dat)" from the List Files of Type box.

(4) Select the appropriate Drive location (e.g., c:) from the Drives box.

(5) The current directory is displayed above the list of directories (this is also represented as the last/lowest open file folder icon in the list). To change the directory location, double-click on the name of the directory from the directories list in the box.

Figure 1.3 Read ASCII Data File Dialog Box

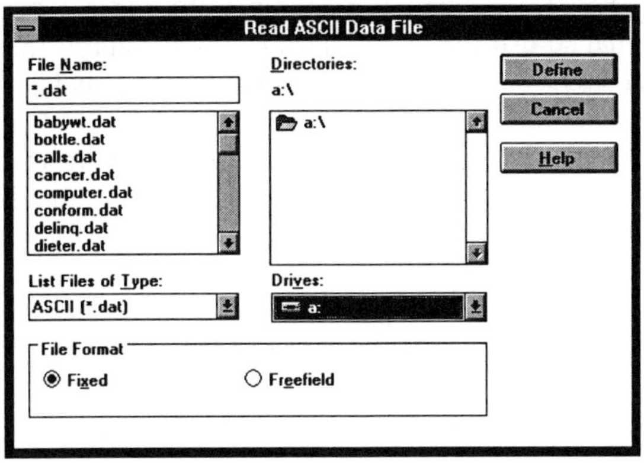

(6) Select the filename either by typing the name of the ASCII file in the File Name box, or else click on the name of the file from the list of filenames that appear in the box under File Name. For example, to read in the data file "dieter.dat," click on that file name.

(7) Select **Fixed** from the File Format box. This indicates that each variable appears in the same column location on the same line for each case.

Once you have told SPSS that you will be reading a fixed format ASCII file, it will want to know some information about the variables in the file. To give this information, you first need to name each variable and then provide additional information as follows:

(1) Click on **Define** in the Read ASCII Data File dialog box. This will open the Define Fixed

Variables dialog box.

(2) Type in the name of the variable ("weight") in the Name box.

(3) Indicate the line (record) in which the variable appears. If there is only one line (row) of data per case -- the most common situation -- then the record number is 1. If there is more than one line of data per case, you need to indicate which line number the variable appears on in the Record box. Since there is only one line of data per case in the "dieter.dat" file, enter 1.

(4) Type the column number (in this example, 1) in which the variable begins in the Start Column box.

(5) Type the column number in which the variable ends in the End Column box. For example, a four-digit variable may begin in column 3 and end in column 6. In the dieter example, the "weight" variable ends in column 3.

(6) Choose the appropriate Data Type. If the numbers in the data file should be read as they appear, select **Numeric as is**. For example, 345 is read as 345, or 1.234 is read as 1.234. Alternate data types are available, such as when the decimal point is not explicity typed in the data file, but you wish to have SPSS insert one. For example, **Numeric 1 decimal** reads the value and places an implied decimal so that one value appears to the right of the decimal point. For example, 990 would be read as 99.0. In this case, select **Numeric as is**.

(7) Blank cells are treated as missing data by clicking on **System missing** in the box labeled "Values assigned to blanks for numeric variables." If you wish blanks in the data to represent a value such as zero or some other number, click on **Value** and enter the value in the box. There are no missing values in the "dieter.dat" file.

(8) Click on the **Add** button when you have completed the variable definition.

(9) Once all of the variable definitions in your data file have been entered, click on **OK**.

1.2.2 Entering Your Own Data

Raw data may be entered in SPSS by using the SPSS data editor. (ASCII data may also be entered with another editor, which are then read by SPSS as described in the previous section.) The SPSS editor looks like a spreadsheet or grid and is automatically opened each time you start an SPSS session. The editor is used to enter, edit, and view the contents of your data file. If you are opening an existing data file (as demonstrated above), the data will appear in the editor and you may then use the editor to change the data or add or delete cases or variables. If you are starting from scratch and wish to enter data, the data editor will be empty when it is first opened.

The data editor is a rectangle defined by rows and columns (see the innermost window in Figure 1.1). Each cell represents a particular row by column intersection (e.g., row 1, column 3). All data files in the data editor have a standard format. Each row of the editor represents a case (e.g., subject #1 or John Doe). Each column in the editor represents a variable (e.g., heart rate or gender). Cells in the editor may not be empty. That is, if the variable is numeric and there is no valid value, the cell is represented by a "system-missing" value and a period appears in the cell. If the variable is a string variable, a blank cell is considered to be valid. (See Section 1.6 for further information on the treatment of missing values.)

To begin entering data in the data editor, follow these steps:

(1) Click on the cell in which you wish to enter data (or use the arrow keys to highlight the cell). Begin at the uppermost left cell in the rectangle. This is row 1, column 1. Once you have clicked on that cell, a darkened border will appear around the cell; this tells you that this is the cell you have selected.

(2) Type in the value you wish to appear in that cell and then press enter. You should notice that the value you type will first appear at the top of the data editor window, and then is inserted in the cell when you press enter. Notice that by entering a value in this first column, you will have automatically created a variable with the default name VAR00001 and it appears at the top of the column. Later we will show how to specify original names and alternate formats for each variable. As an example, suppose that you are recording ages for 25 people. If the age of the first person is 18, enter 18 in the first cell.

(3) Type in another value. Again, you will see it at the top of the data editor, and then it will appear in the next cell. The next cell will be directly below the previous cell. This location will be row 2, column 1. Suppose the age of the second person was 22, enter 22 in row 2 column 1.

(4) Repeat this process until you have entered all of the data you wish for column 1 (values for all cases on variable 1).

(5) Now click on the first cell in the next column (row 1, column 2). This will automatically create a new variable and call it VAR00002. Suppose that "shoe size" is the next variable, and the first person has size 7.

(6) Type in the value for the first cell in column 2 and press enter. Enter 7.

(7) Repeat this process for all values in column 2.

(8) Continue this procedure until you have entered values for all cases and variables that you wish for your data file.

Once you have entered data in the data editor, you may change or delete values. To change or delete a value in a cell, simply click on the cell you wish to alter. You will notice that a dark border appears around the selected cell, and the value in the cell appears at the top of the data editor. If you are changing the value, simply type the new value and press enter. You should see the new value replace the old value in the cell.

Adding Cases and Variables

To insert a new case (row) in between cases that already exist in your data file:

(1) Point the mouse arrow and click on the row *below* the row where you wish to enter the new case.

(2) Click on **Data** on the menu bar.

(3) Click on **Insert Case** from the pull-down menu.

A new row is now inserted and you may begin entering data in the cells. Notice that before you enter your values, all of the cells have system-missing values (represented by a period).

To insert a new variable (column) between existing variables:

(1) Click on the column that is to the *right* of the position where you wish to enter a new variable.

(2) Click on **Data** on the menu bar.

(3) Click on **Insert Variable** from the pull-down menu.

A new variable (column) is now inserted and you may begin entering data in the cells.

Deleting Cases and Variables

To delete a case:

(1) Click on the case number (on the left side of the row) that you wish to delete.

(2) Click on **Edit** from the menu.

(3) Click on **Clear**.

The selected case will be deleted and the rows below will shift upward.

To delete a variable:

(1) Click on the variable name (on the top of the column) that you wish to delete.

(2) Click on **Edit** from the menu.

(3) Click on **Clear**.

The selected variable will be deleted and all variables to the right of the deleted variable will shift to the left.

Defining Variables

By default, SPSS assigns variable names and formats to the variables in the SPSS data file. By default, variables are named VAR##### (prefix VAR followed by five digits) and all values are valid (blanks are assigned system-missing values). Most of the time, however, you will want to customize your data file. For example, you may want to give your variables more meaningful names, provide labels for specific values, change the variable formats, and assign specific values to be regarded as "missing." To do any or all of these:

(1) First, make sure that your data file window is the active window and click on the variable that you wish to change.

(2) Click on **Data** in the menu bar.

(3) From the Data pull-down menu, click on **Define Variable**. This opens the Define Variable dialog box (see Figure 1.4).

Figure 1.4 Define Variable Dialog Box

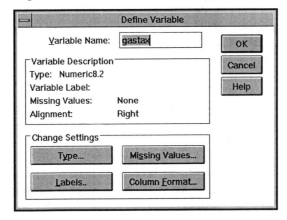

(4) The name of the variable will appear in the Variable Name box. To change it, simply type the new name in this box. The name has to begin with a letter, cannot contain blanks, and cannot exceed 8 characters. If this is all you wish to do, click on **OK**.

(5) If you wish to change the type or format of a variable (in the Data Editor window and outputs), click on **Type** to open the Define Variable Type dialog box. In most cases you will not need change the variable format. One situation in which you might wish to change the format is as follows: Suppose the variable representing average cost of groceries per person was entered to the nearest cent (e.g., 32.24). If you wish to change this format so that the average cost is displayed as a whole number (rounded to the nearest dollar, e.g., 32) you would use this option. By default, all variables are numeric. Click any other type listed below if you want the variable to be anything other than numeric.

If you wish to change the format of the numeric variable, click in the **Width** box. The number in this box tells you the total number of columns that the variable occupies in the data file (including one column for decimal places, plus, or minus signs). For example, 8 indicates that the variable is 8 columns long. Type in the variable's column width. If you wish to change the number of decimal places, click in the **Decimal Places** box. The number in this box tells you how many numbers appear after the decimal place. For example, the number 32.24 would have a "width" of 5 and a 2 in the "decimal places" box. The number 32 would have a width of 2 and a 0 in the decimal places box. Click **Continue** when you are finished.

(6) If there are specific values that you would like to be treated as missing values, click on **Missing Values** to open the Define Missing Values dialog box. Click on **Discrete Missing Values** to tell SPSS that you have specific values that are considered to be missing. Type in the value(s) in the boxes (you may add up to three values). If you have more than three missing values, click on **Range** of missing values and enter the lower and upper bounds. Click **Continue** when you have entered in all of your missing values.

(7) If one of your variables is categorical, SPSS will automatically assign numbers representing the categories of the variable. For example, the variable gender will have 2 categories: male and female. Males may have the assigned value "1" and "2" will represent females. It is useful to have alphabetic labels assigned to the values of 1 and 2 so that it is easy to see which number represents which category in your output files.

To assign values labels to the variable, click on **Labels**. This opens the Define Labels dialog box. First type the name of variable (e.g., gender) if it does not already appear in the variable label box. Tab to, or click on the **Value** box. Type the number representing the first category (e.g., 1). Go to the Value Label box and type the label for this value (e.g., male). Now click on the **Add** button. Go back to the Value box and type in the next value (e.g., 2). Type the value label for this value in the Value Label box (e.g., female), and click on **Add**. Note that each time you click Add, you will see the value and it's corresponding label appear in the window to the right of the Add button. When you have added all of the values and labels, click on **Continue**.

1.2.3 Saving Files

Saving SPSS Data Files

Unless you save your files, all of your data and data information will be lost when you leave the SPSS session. To save a file, first make the Data Editor the active window. Then:

(1) Click on **File** from the menu.

(2) Select **Save Data** from the list of options in the File pull-down menu.

(3) Type the name of your file in the **File Name** box in the upper-left corner of the window. Notice that it is automatically set for any filename with the ".sav" extension.

(4) Click **OK**.

By default, this will save the data file as an SPSS data file. If you were working with a previously existing data file, the old file will be overwritten by the modified data file.

Saving ASCII Data Files

If you are saving an alternate type of data file, such as an ASCII file, the procedures are slightly different.

(1) With the Data Editor as the active window, click on **File** from the menu.

(2) Click on **Save Data As** from the File pull-down menu.

(3) Choose the appropriate Drive and Directory.

(4) Select **Fixed ASCII (*.dat)** from the Save File as Type box.

(5) Type the name of your file in the File Name box. Note that the file will be set for any filename you choose with the ".dat" extension.

(6) Click on **OK**.

1.3 Transforming Variables and Data Files

At times, you may need to alter or transform the data in your data file to allow you to perform the calculations you require. There are many ways in which you can transform data.

This section discusses three commonly used techniques: computing new variables, recoding variables, and selecting subsets of cases.

1.3.1 Computing New Variables

There may be occasions when you will need to compute new variables that combine or alter existing variables in your data file. For instance, your data file may contain SAT-verbal and SAT-math scores for a group of high school seniors, but you are interested in examining total SAT scores.

To create a new variable:

(1) Click on **Transform** from the menu bar.

(2) Click on **Compute** from the pull-down menu. This opens the Compute Variable dialog box (see Figure 1.5).

Figure 1.5 Compute Variable Dialog Box

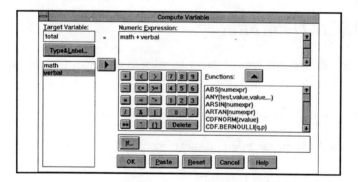

3) Enter the name of the new variable (in the above illustration, total) in the Target Variable box. (You also have the option to describe the nature and format of the new variable by clicking on the **Type & Label** box.)

(4) You will then need to perform a series of steps to construct an expression used to compute your new variable. In the SAT illustration, you would first select the math variable from the Variable List box on the lefthand side of the dialog box and move it to the Numeric Expression box using the **right directional arrow.**

(5) Then click on the "+" from the calculator pad. You will notice that a plus sign is placed in the Numeric Expression box after the word math.

(6) You would then complete the expression by selecting the verbal variable and moving it to

the Numeric Expression box, following the instructions in step (4) above.

(7) When you have completed the expression, click on **OK** to close the Compute Variable dialog box. You will see the message "Running Execute" at the bottom of the application window indicating that SPSS is computing the new variable. When computations are complete, this message will be replaced with "SPSS Processor is Ready" and your new variable will appear in the first empty column in the data editor window.

In addition to simple algebraic functions on the calculator pad (+, −, x, ÷), there are many other arithmetic functions such as absolute value, truncate, round, square root, and statistical functions including sum, mean, minimum, and maximum. These are displayed in the function box to the right of the calculator pad.

1.3.2 Recoding Variables

Recoding variables is often a useful technique. There are several instances in which you may need to change the values in your original data file. You can recode either categorical or numeric variables. For instance, you may have a data file containing population size, income, and homeowner information for indiviudals based on the state in which they live. That is, there would be one state variable, with a value of 1 through 50, indicating the subject's home state. Suppose, however, that you are interested in analyzing income distribution based on whether individuals live in the Eastern or the Western portion of the country. For this, you would need to recode the state variable.

Or, you may want to categorize a discrete or continuous numeric variable into a limited number of groups. For example, your data file may contain ages of 200 individuals, and you wish to group them into young (for your purposes, under age 40), middle (ages 40-64) and old (over 64) categories.

You have two options available for recoding variables. You may recode values into the same variable, which eliminates all record of the original values. You also have the option to create a new variable containing the recoded values. This preserves the original values of the variable. If you think that there may be a reason that you would need to have record of the original values, you should select the second option.

Recoding into the Same Variable

To recode into the same variable:

(1) Click on **Transform** from the main menu.

(2) Click on **Recode >** from the pull-down menu.

(3) Click on **Into Same Variable** from the supplementary menu that appears after you choose Recode. This opens the Recode into Same Variable dialog box.

(4) Select the name of the variable to be recoded, and move it to the Variables box with the **right arrow button**.

(5) Click on **Old and New Values**. This opens the Old and New Variables dialog box (see Figure 1.6).

(6) For each value (or range of values) you want to recode, you must indicate the old value and the new values, and click on **Add**. The recode expression will appear in the **Old --> New** box.

Figure 1.6 Recode Into Same Variables: Old and New Values Dialog Box

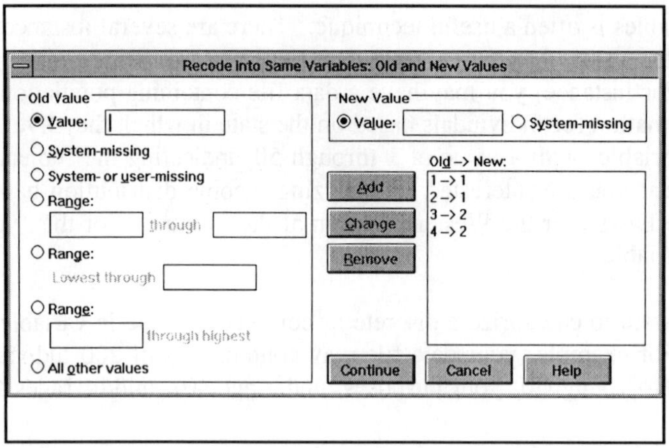

Old values are those before recoding, the values that exist in the original data file. There are several alternatives for old values, including the Value and the Range options discussed below.

Value Option. You may use the Value option in cases in which you want to collapse existing categorical variables such as the "state" illustration above. In your original data file, you have a separate code (1 through 50) for each state, but you would recode this to a variable with 1 representing Eastern states and 2 representing Western states. For example:

Original		Recoded	
Category	Value	Category	Value
NY	1	Eastern	1
PA	2	Eastern	1
CA	3	Western	2
AZ	4	Western	2

The steps to perform this recoding are straighforward:

(1) Type **1** in the **Value** box of the **Old Value** section, indicating State 1.

(2) Type a **1** in the **Value** box of the **New Value** section, indicating that it is to be recoded as region 1 (Eastern).

(3) Click on **Add**. You will notice that the expression 1 --> 1 will appear in the **Old -->New** box.

Follow the same procedure for the rest of the values of the original variable. For example, you would recode the old value of 2 to the new value of 1, and click on **Add**. Note that because NY was coded 1 both before and after recoding, it was not necessary to include it in the recode procedure. Doing so, however, may assist you in making sure that you have included all values to be recoded.

Range Option. You may also recode variables using the Range option. This is most useful for numerical variables. The procedure is similar to that discussed above. In the age illustration, values would be recoded as follows:

Original	Recoded	
Range of Values	Category	Value
Lowest through 39	Young	1
40 through 64	Middle	2
65 through Highest	Old	3

Instead of chosing the value option in the old value section, you may use the range option as follows:

(1) Type **39** in the **Lowest through** ___ box, the second of the range options.

(2) Type **1** in the **Value** box under the New Values section.

(3) Click on **Add**. Again, the expression will appear in the **Old --> New** box.

(4) Type **40** and **64** in the two boxes of the first range option: ___ **through** ___.

(5) Type **2** in the **Value** box under the New Values section.

(6) Click on **Add**.

(7) Type **65** in the empty box in the ___ **through Highest** range option of the Old Values section.

(8) Click on **Add**.

When you have indicated all the recode instructions, using either the Value or Range method, click on **Continue** to close the Recode Into Same Variables: Old and New Values dialog box. Click on **OK** to close the Recode Into Same Variables dialog box. While SPSS performs the transformation, the message "Running Execute" appears at the bottom of the application window. The "SPSS Processor is Ready" message appears when transformations are complete.

Recoding into Different Variables

The procedure for recoding into a different variable is very similar to that for recoding into the same variable:

(1) Click on **Transform** from the main menu.

(2) Click on **Recode >** from the pull-down menu.

(3) Click on **Into Different Variable** from the supplementary menu that appears after you choose Recode. This opens the Recode into Different Variable dialog box.

(4) Select the name of the variable to be recoded from the variables list, and move it to the Input Variables --> Output Variable box with the **right arrow button**.

(5) Type in the name of the variable you wish to create in the Name box of the Output Variable. If you wish, you may also type a label for the variable.

(6) Click on **Change**, and the the new variable name will appear linked to the original variable

in the Input Variable --> Output Variable box.

(7) Click on **Old and New Values**. This opens the Old and New Variables dialog box.

(8) The procedure for determining old and new values is the same as that discussed in the Recoding into the Same Variable subsection, with one exception. Because you are creating a new variable, you must indicate new values for *all* of the old values, even if the value does not change. (This is optional when recoding to the same variable.) Because this step is mandatory, SPSS provides a **Copy Old Value(s)** option in the New Value box.

(9) When you have indicated all the recode instructions, click on **Continue** to close the Recode Into Same Variables: Old and New Values dialog box.

(10) Click on **OK** to close the Recode Into Same Variables dialog box. While SPSS performs the transformation, the message "Running Execute" appears at the bottom of the Application Window. The "SPSS Processor is Ready" message appears when transformations are complete, and a new variable appears in the data editor window.

1.3.3 Selecting Cases

There may be occasions in which you may need to select a subset of cases from your data file for a particular analysis. You may, for instance, have a data file containing heights and weights of 200 individuals, but you need to know the average height of individuals over 120 pounds. Or, you may simply wish to choose a random sample of cases in a very large data file.

To select subset of cases:

(1) Click on **Data** from the main menu.

(2) Click on **Select Cases** from the pull-down menu. This opens the Select Cases dialog box (see Figure 1.7).

There are several ways in which you can select a subset of a larger data file. We will discuss the If Condition and Sample methods.

If Condition

One of the methods of selecting subcases is through the If Condition option. In the height and weight example given above, you would need to:

(1) Select the **If Condition is Satisfied** option and click on **If** to open the Select Cases: If dialog box.

(2) Select the weight variable from the variable list box, and move it into the box above the calculator pad with the **right arrow button**.

(3) Using the calculator pad, click on >. This sign will appear in the upper box.

(4) Still using the number pad, Click on **1**, then **2**, then **0**, to create the expression "weight > 120" in the upper right-hand box.

(5) Click on **Continue** to close the Select Cases: If dialog box.

Figure 1.7 Select Cases Dialog Box

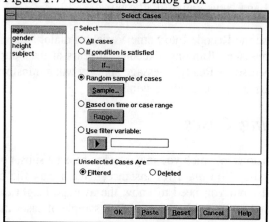

Random Sample

Another method for selecting subcases is to choose a random sample from your data file:

(1) From the Select Cases dialog box (see Figure 1.7, above) select the **Random Sample of Cases** option and click on **Sample** to open the Select Cases: Random Sample dialog box.

(2) Type either a percentage or a precise number of cases in the appropriate box.

(3) Click on **Continue** to close the Select Cases: Random Sample dialog box.

You should now be back at the Select Cases dialog box. Click on **OK** to close this dialog box. The message "Running Execute" will appear as SPSS processes your instructions. SPSS creates a new "filter" variable with the value 1 for all cases selected and 0 for all cases not selected. There is also a "Filter On" message at the bottom of the application window to remind you that your subsequent analyses will be performed only on that subset of data.

Furthermore, there are diagonal lines through the cell line numbers of the data window for all cases not selected. SPSS uses this variable to determine which cases to include in subsequent analyses. You do not need to refer to it in any of your commands, since this is simply a code within the SPSS program.

To turn the filter off, go back into the Select Cases dialog box and click on **All Cases** in the Select box, and then click on **OK**.

1.4 Examining and Printing Output

SPSS provides two different types of output windows (see Section 1.1.1). The basic Output Window contains text-based results of most statistical analyses, including descriptive statistics, frequency distributions, results of t-tests, and the like. The Chart Carousel Window, on the other hand, contains charts and plots such as histograms, box-and-whisker plots, and scatterplots.

1.4.1 Output Window

The Output Window opens automatically each time you start SPSS for Windows. Each time you direct SPSS to perform a statistical procedure, the results are automatically added to the Output Window. New results follow previous results, and the cursor moves down to the last line of output. You may view the output on the screen by using the directional arrows buttons on the vertical scroll bar at the right edge of the window.

1.4.2 Chart Window

Each time you direct SPSS to create a chart or graph, it creates a new Chart Carousel Window with the graph. Chart windows are numbered consecutively. The menu and icon bar for the Chart Window contain options for editing charts. For instance, you are able to create three-dimensional charts, to change chart colors, to create and edit chart and axis titles, and the like. Because these options are mainly for presentation, but do not affect the substantive information contained in the charts, they are not discussed in detail here. If you have trouble changing attributes of your charts, you may consult the Help facility.

1.4.3 Printing Output

To print the contents of an Output Window:

(1) Make the Output Window from which you wish to print the active window (see Section 1.2).

(2) Click on **File** from the main menu.

(3) Click on **Print** from the pull-down menu. This opens the Print dialog box.

(4) If you wish to print the entire file, click on the **All** option. if, on the other hand, you only wish to print a selected block, click on the **Selection** option. To print only a section of the file, you need to use the "click-and-drag" method to highlight the area before opening the Print dialog box.

(5) Click on **OK** to close the Print dialog box and print the file.

1.5 Missing Values

In many situations, data files do not have complete data on all variables; that is, there are missing values. You need to inform SPSS when you have missing values so that all computations are performed correctly. With SPSS, there are two forms of missing values: system-missing and user-defined missing.

System-missing values are those that SPSS automatically treats as missing (without the user having to explicitly inform SPSS). The most common form of this type of value is when there is a "blank" in the data file. For example, a value for a variable may not be entered in the data file if the information was not provided. When SPSS reads this variable, it will read a blank, and thus treat the value as though it is missing. Any further computations involving this variable will proceed without the missing information. For instance, suppose you wish to calculate the average amount of daily water consumption for a sample of 20 adults, but you only have data entered for 19 people. SPSS will read the "valid" values for the 19 adults, ignore the missing value, and compute the average based on the 19 individuals. System-missing values are represented by a period in the SPSS data editor and in the output of some procedures.

User-defined missing values are those that the user specifically informs SPSS to treat as missing. Rather than leaving a blank in the data file, numbers are often entered that are meant to represent data. For example, suppose you have an age variable that ranges from 1 through 85. You could use the number 99 to represent those cases that were missing information on age. (You could not use any numbers from 1-85 since these are all valid values.)

You need to inform SPSS that 99 is to be treated as a missing value, otherwise it will treat is as valid. This is explained in Section 1.2.2, but in brief, you need to do the following: from the Define Variable dialog box (Figure 1.4), click on **Missing Values**, click on **Discrete Missing Values**, type in 99, click on **Continue** (see Section 1.2.2). When SPSS reads this variable, it will then treat 99 as a missing value and not include it in any computations involving the "age" variable. User-missing values will look like valid values in the data editor, but are internally flagged as "missing" in SPSS data files, and labeled as missing in the output for some procedures. Values that are user-defined as missing are displayed in the Define Variable dialog box (Figure 1.4) and labeled as Missing Values.

Most SPSS computations will display the valid number of cases in the output. This is the number of cases that were not system-missing and/or user-defined missing; these values are used in the computations. The number of missing cases (not used in the computations) is typically displayed as well.

Chapter Exercises

1.1 Begin an SPSS session and select the "football.sav" data file. Without opening the file, answer the following:

(a) How many variables are in the file?
(b) What is the format for the weight variable?

1.2 Open the SPSS data file "spit.sav" and answer the following:

(a) Is this an ASCII file?
(b) How many cases are in the file?
(c) How many variables are in the file?
(d) Are there any missing data?

1.3 Enter the data for fast and slow learners in Table 2.24 of the textbook into a new data file using the SPSS data editor. (You will need to create a variable for type of learner.)

(a) Save the data as an SPSS data file.
(b) Save the data as an ASCII data file.
(c) Once you have saved and exited the file, re-open the ASCII data file and enter a new variable named "errors" with the value of 1 for all fast learners and 3 for slow learners.
(d) Re-open the SPSS data file and delete the fifth case. Was this a fast or slow learner?

1.4 Open the "semester.sav" data file and use SPSS to do the following:

(a) Compute a new variable that is the average of the final grades.
(b) Recode the statistics courses variable into a new variable that distinguishes students who either took at least one statistics course or who did not. Create new value labels.
(c) Recode the "major" variable so that there are two majors only: science and non-science. Create new value labels.
(d) Select out cases that have taken 2 statistics courses. How many cases are there?

PART II

DESCRIPTIVE STATISTICS

Chapter 2   ORGANIZATION OF DATA

Section 2.1 of the textbook describes different types of measurement scales. A distinction is made in this chapter between numerical and categorical data. Numerical data are further classified as discrete or continuous. This chapter demonstrates how to examine different types of data through frequency distributions and other graphical representations.

2.2 Organization of Categorical Data

Information on the coding and labeling of categorical data is given in Chapter 1. When numbers have been assigned to the different categories, SPSS uses them as additional labels when constructing frequency distributions.

Frequencies

Let us begin by creating a simple frequency distribution of occupations using the "occup.sav" SPSS data file on the disk provided with this manual. You may follow along by using SPSS to open the data file on your computer (using the procedure given in Chapter 1). This is a data set containing information on the occupations of primary householders as shown in Table 2.1 of the textbook. Notice in the data file that each category is represented by a number (e.g., 1, 2, 3, 4). You may examine the contents of the data file (variable labels, variable type, and value labels) by clicking on **Utilities** on the menu bar and clicking on **Variable Information** from the pull-down menu.

To create a frequency distribution of these data:

(1) Click on **Statistics**.

(2) Click on **Summarize** from the pull-down menu.

(3) Click on **Frequencies** from the second pull-down menu to open the Frequencies dialog box (see Figure 2.1).

(4) Click on the name of the variable that you wish to examine ("occup").

(5) Click on the **right arrow button** to move the variable name into the Variables box.

(6) Click on **OK**.

The frequency distribution produced by SPSS is shown in Figure 2.2.

Figure 2.1 Frequencies Dialog Box

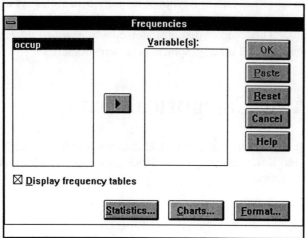

Figure 2.2 Frequency Distribution of Occupation Variable

```
OCCUP       occupation


                                                  Valid      Cum
Value Label                Value  Frequency  Percent  Percent  Percent

professional                 1        6       30.0     30.0     30.0
sales                        2        8       40.0     40.0     70.0
clerical                     3        4       20.0     20.0     90.0
laborer                      4        2       10.0     10.0    100.0
                                   -------  -------  -------
                    Total            20      100.0    100.0


Valid cases      20    Missing cases      0
```

The variable categories appear in the column labeled "Value Label." The "Value" column contains the numbers that represent the category (note that these are the numbers that appear in the data file). The "Frequency" column contains that exact number of cases (e.g., number of householders) for each of the categories. For example, there are 6 professional householders and 4 clerical householders.

The numbers in the "Percent" column represent the percentage of the total that is contained each category. They are obtained by dividing frequency by the total number of cases. For example, 30% of the entire sample is professional householders (100 x 6/20).

The "Valid Percent" column takes into account missing values. For instance, if there were two missing values in this data set, then the valid number of cases would be 18. If that were the case, the Valid Percent of professional householders would be 33%. Note that "Percent" and "Valid Percent" will both always total to 100%.

The "Cumulative Percent" is a cumulative percentage of the cases for the category and all categories listed above it in the table. For example, 90% of all cases in the sample include professional, sales, and clerical types of householders (i.e., 30% + 40% + 20%). The cumulative percentages are not meaningful, of course, unless the scale has ordinal properties, which this variable does not. That is, the occupations could have been listed in another order without affecting the interpretation of the data.

Bar Charts

A bar chart is also useful for examining categorical data. In a bar chart, the height of each bar represents the frequency of the occurrence for each category of the variable. Let us create a bar chart for the occupation data using an option within the Frequencies procedure.

From the Frequencies dialog box:

(1) Click on **Chart** to open the Frequencies: Charts dialog box (see Figure 2.3).

(2) Click on **Bar chart** in the Chart Type box.

(3) Click on **Continue**.

(4) Click on **OK** to close the Frequencies dialog box and run the chart procedure.

(5) SPSS puts the bar chart in a Chart Carousel window. To open this window, double click on the **Chart Carousel** icon in the lower left-hand corner of the screen. A bar chart like that in Figure 2.4 should appear on your screen.

Figure 2.3 Frequencies: Charts Dialog Box

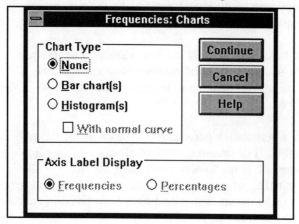

Figure 2.4 Bar Chart of Occupation Variable

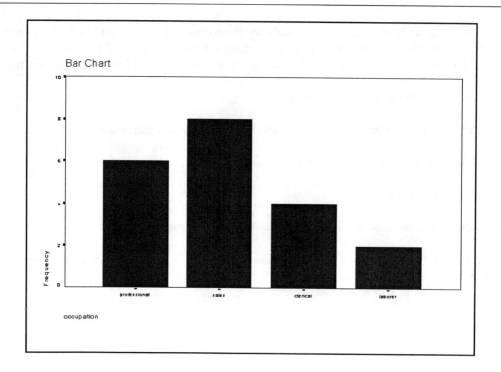

Notice that the occupation with the greatest number of people is sales, while the fewest number of people are laborers. There are six professionals which can be determined by looking at the vertical axis (frequency).

2.3 Organization of Numerical Data

Discrete Data

You can also obtain frequency distributions and graphical representations for discrete numerical variables. Using the data from Table 2.6 of the textbook (contained in the data file "kids.sav"), let us create a frequency distribution of the number of children in a household using the same procedures as outlined in Section 2.2.1 above.

A histogram is a graphical procedure often used to display numerical data. The procedure is identical to the one used to obtain a bar chart in Section 2.2.1 except you need to click on **Histogram** (instead of Bar Chart) in the Frequencies: Charts dialog box. Again, you will need to double click on the Chart Carousel icon in order to see the histogram shown in Figure 2.5.

Figure 2.5 Histogram of Number of Children in Household

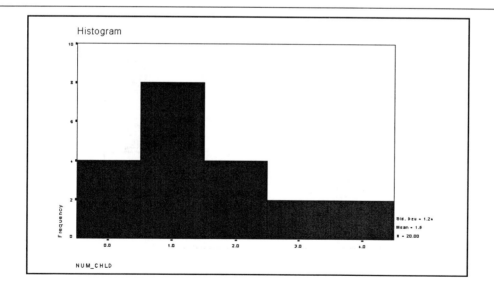

The horizontal axis represents the number of children in the household, with values from 0 to 4. The vertical axis represents the number of occurrences of each value of x. For example, there are four occurrences of the value zero, which means that there are four households with no children.

Frequency Distributions for Continuous Data

We shall use the data in Table 2.10 of the textbook on the weights of 25 dieters to obtain a histogram and a stem-and-leaf plot; you may follow along by first opening the file "dieter.sav" on the diskette. Since the data are continuous, we use an alternative method for creating histograms, as follows:

(1) Click on **Statistics** on the menu bar.

(2) Click on **Summarize** from the pull-down menu.

(3) Click on **Explore** from the pull-down menu. This opens the Explore dialog box as shown in Figure 2.6.

Figure 2.6 Explore Dialog Box

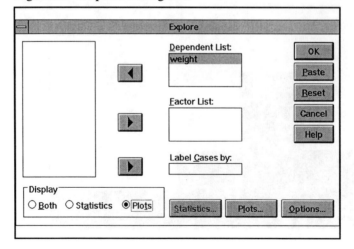

(4) Click on the name of the variable ("weight") and click on the **top right arrow button** to move it to the Dependent List box.

(5) Click on **Plots** in the Display box. This will suppress all statistics in the output. If you also

want SPSS to provide summary statistics, click on **Both**.

(6) Click on the **Plots button** to open the Explore: Plots dialog box.

(7) Click on **Histogram** in the Descriptive box.

(8) Click on **Continue**.

(9) Click on **OK** to run the procedure.

(10) Double click on the Chart Carousel icon to view the histogram shown in Figure 2.7.

Figure 2.7 Histogram of Weight of Dieters

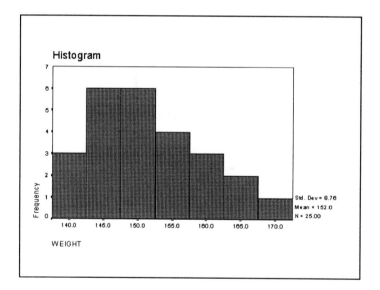

Changing Intervals

Since the "weight" variable has a range of 34 pounds, the histogram with 7 intervals (selected by SPSS) is adequate. It is possible to edit the histogram to change the number of intervals displayed or the interval width. For example, to change the number of intervals in the above histogram to 10:

(1) In the Chart Carousel window, click on **Edit** to open the Chart window.

(2) Click on **Chart** from the menu bar.

(3) Click on **Axis** from the pull-down menu to open the Axis Selection dialog box.

(4) Click on **Interval.**

(5) Click on **OK** to open the Interval Axis dialog box (Figure 2.8).

Figure 2.8 Interval Axis Dialog Box

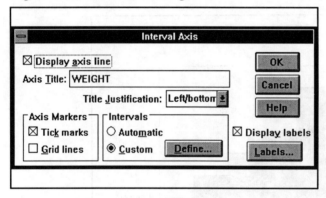

(6) In the Intervals box, click on **Custom**.

(7) Click on the **Define** button to open the Interval Axis: Define Custom Intervals dialog box (Figure 2.9).

(8) In the definition box, change the "# of intervals" from 7 to 10.

(9) Click on **Continue** to close this dialog box.

(10) Click on **OK** to close the Interval Axis dialog box and re-draw the graph. (See Figure 2.10).

Figure 2.9 Interval Axis: Define Custom Intervals Dialog Box

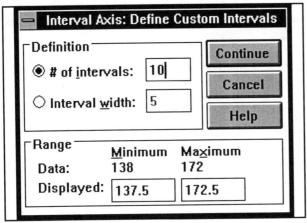

Figure 2.10 Histogram of Weight of Dieters (10 Intervals)

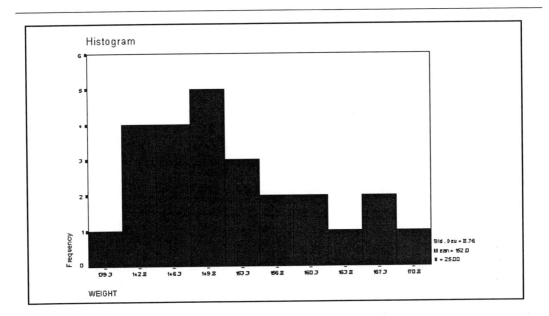

Stem-and-Leaf Plot

SPSS will also produce a stem-and-leaf display of the data. To accomplish this, follow the steps 1-6 given above, plus:

(1) Click on **Stem-and-leaf** in the Descriptive box of the Explore: Plots dialog box.

(2) Click on **Continue**.

(3) Click on **OK** to run the procedure.

The stem-and-leaf plot will appear in the output window, as shown in Figure 2.11.

Figure 2.11 Stem-and-Leaf Display of Weight

```
    WEIGHT

Valid cases:         25.0  Missing cases:        .0  Percent missing:      .0

Frequency     Stem &  Leaf

     1.00      13 .   8
     4.00      14 *   1234
     7.00      14 .   5677899
     5.00      15 *   01234
     2.00      15 .   57
     3.00      16 *   012
     2.00      16 .   67
     1.00      17 *   2

Stem width:          10
Each leaf:            1 case(s)
```

The stem-and-leaf plot is similar to a histogram. Notice that the weights range from 138 [(13 x 10) + 8] to 172 [(17 x 10) + 2] pounds. The frequency column gives the number of observations in each interval. Notice that there are two sections of stem with the value of 14. The asterisk after the first of these stems indicates that all observations ranging from 140 to 144 would be in this interval. Observations ranging from 145 to 149 are listed in the interval marked with a period to the right of the second 14.

Chapter Exercises

2.1 Open the "football.sav" data file containing heights and weights of football players from Stanford University (Table 2.15), and use SPSS to perform the following analyses:

(a) Create a frequency distribution of weights. How many players weigh 195 pounds?
(b) What percent of players weigh 200 pounds or less?
(c) What is the highest and lowest weight?
(d) Create a histogram of the heights of the players.
(e) What is the midpoint of the interval with the highest frequency? How many players are in this height interval?

2.2 Open the "fire.sav" data file (Table 2.22) which contains demographic and performance data on 28 firefighter candidates, and use SPSS to answer the following:

(a) How many male firefighter applicants are there in the sample? What percentage of the total number of applicants is this?
(b) What percent of applicants are minority group members?
(c) For males only, what written test score occurs most frequently? (Hint: you need to use the Select If command detailed in Chapter 1.)
(d) What percent of females scored below 18 seconds on the stair climb task?
(e) Create a stem-and-leaf plot for the written test score.
(f) How many applicants had a score between 85 and 89?
(g) What is the range of the scores?

2.3 Open the "semester.sav" data file (Table 2.21) on 24 students in a statistics course, and use SPSS to perform the following analyses:

(a) Create a bar chart of the majors. What is the most popular major?
(b) How many students have had 1 semester or fewer of statistics?

Appendix II
SPSS Syntax for Organization of Data

2.2.1 Frequencies

In order to obtain a frequency distribution for the "occup" variable and a bar chart, the SPSS syntax is:

Frequencies variables=occup
/barchart.

2.3.1 Discrete Data

To create a histogram and frequency distribution of the "num_chld" variable, use the following SPSS command:

Frequencies variables=num_chld
/histogram.

2.3.3 Frequency Distributions for Continuous Data

To create a frequency distribution and histogram for the continuous variable "weight," use the following SPSS command:

Frequencies variables=weight
/histogram.

To create a stem-and-leaf plot for the "weight" variable, use the following SPSS command:

Examine variables=weight
/plot=stemleaf.

Chapter 3 MEASURES OF LOCATION

This chapter describes various ways to use SPSS to obtain the mode, the median, and the mean of a sample of data.

3.1 The Mode

The mode, the most common value of a variable, is especially useful in summarizing categorical or discrete numerical data. The easiest way to obtain the mode with SPSS for Windows is by using the Frequencies procedure. This is the same procedure used to obtain frequency distributions, histograms, and bar charts discussed in Chapter 2.

We will illustrate how to obtain the mode using the "semester.sav" data file from Table 2.21 of the textbook, containing information on 24 students in a statistics course. The "major" variable is a categorical variable representing the type of program in which each student is enrolled. To obtain the mode of this variable:

(1) Click on **Statistics** from the menu bar.

(2) Click on **Summarize** from the pull-down menu.

(3) Click on **Frequencies** from the pull-down menu.

(4) Click on the **right arrow button** to move the "major" variable into the Variables box.

(5) Click on the **Statistics button** at the bottom of the screen. This opens the Frequencies: Statistics dialog box, as shown in Figure 3.1.

Figure 3.1 Frequencies: Statistics Dialog Box

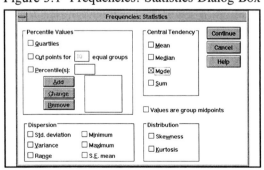

(6) Locate and click on the **Mode** option in the Central Tendency box.

(7) Click on **Continue** to close this dialog box.

(8) Click on **OK** to close the Frequencies dialog box and execute the procedure.

The output will look similar to that discussed in Chapter 2 and shown in Figure 3.2.

Figure 3.2 Frequency Distribution with Mode for College Major

```
MAJOR

                                                      Valid       Cum
Value Label              Value   Frequency  Percent   Percent    Percent

electrical engineer      1.00         2       8.3       8.3        8.3
chemical engineer        2.00         4      16.7      16.7       25.0
statistics               3.00         2       8.3       8.3       33.3
psychology               4.00         1       4.2       4.2       37.5
public administratio     5.00        11      45.8      45.8       83.3
architecture             6.00         2       8.3       8.3       91.7
industrial administr     7.00         1       4.2       4.2       95.8
materials science        8.00         1       4.2       4.2      100.0
                                 -------   -------   -------
                        Total        24     100.0     100.0

Mode            5.000

Valid cases        24      Missing cases       0
```

Notice that in addition to the frequency distribution, the output lists the mode of the variable; it is occupation 5. In other words, more students in this sample are public administration majors in comparison to all other majors. Since this scale does not have ordinal properties, the cumulative percentages are meaningless.

Mode of Grouped Data

It is also possible to determine the mode of a variable by examining the frequency distribution. As an example, refer back to Figure 3.2. Even without the Mode option, you could search through the frequency column for the row with the highest number, here 11. Or, you could look at the percent column for the largest percentage, here 45.8%. The value associated with these numbers is the most common value for the variable -- the mode of the variable.

3.2 The Median and Other Percentiles

3.2.1 The Median

The median of a variable is the value that divides the distribution in half. The procedure for determining the median of a variable is similar to that for obtaining the mode. You simply need to click on the **Median** option instead of the Mode option in the Central Tendency box of the Frequencies: Statistics dialog box. (See Figure 3.1.)

As an example, find the median of the dieters' weights using the "dieter.sav" data file. This file contains the weights for 25 individuals. Your output should look like that in Figure 3.3. The median of the distribution is 150 pounds; it is indicated at the bottom of the table. Twelve observations have weights below 150 pounds and 12 have weights above 150 pounds.

Figure 3.3 Frequency Distribution with Median for Weight of Dieters

```
WEIGHT
                                                Valid     Cum
Value Label              Value  Frequency  Percent  Percent  Percent

                          138        1       4.0      4.0      4.0
                          141        1       4.0      4.0      8.0
                          142        1       4.0      4.0     12.0
                          143        1       4.0      4.0     16.0
                          144        1       4.0      4.0     20.0
                          145        1       4.0      4.0     24.0
                          146        1       4.0      4.0     28.0
                          147        2       8.0      8.0     36.0
                          148        1       4.0      4.0     40.0
                          149        2       8.0      8.0     48.0
                          150        1       4.0      4.0     52.0
                          151        1       4.0      4.0     56.0
                          152        1       4.0      4.0     60.0
                          153        1       4.0      4.0     64.0
                          154        1       4.0      4.0     68.0
                          155        1       4.0      4.0     72.0
                          157        1       4.0      4.0     76.0
                          160        1       4.0      4.0     80.0
                          161        1       4.0      4.0     84.0
                          162        1       4.0      4.0     88.0
                          166        1       4.0      4.0     92.0
                          167        1       4.0      4.0     96.0
                          172        1       4.0      4.0    100.0
                                  -------  -------  -------
                          Total     25      100.0    100.0

Median         150.000

Valid cases       25      Missing cases       0
```

As with the mode, you can also determine the median from the frequency distribution. To do so, start at the top and read downward until you locate the first row in the distribution that has a cumulative greater than 50 percent. The value associated with this percentage divides the distribution in half; it is the median. Here, 48% of the individuals weigh 149 pounds or less, and 52% weigh 150 pounds or less. Again, 150 pounds is the median of this distribution.

If a value in a frequency distribution has a cumulative percentage that is exactly 50, then the median is halfway between that value and the subsequent value. For example, if the cumulative percent for 149 pounds were 50%, then the median of the distribution of weights would be (149 + 150)/2 = 149.5 pounds.

If you would like to obtain both the mode and the median of a variable, you can select more than one option from the Frequencies: Statistics dialog box and obtain several statistics at once. There may also be times that you wish to only obtain the statistics, but not the frequency distribution. (This will be useful for examining continuous variables from very large data sets.) Suppose you have a data file containing heights of 500 people. If the heights were measured to the nearest one-tenth of an inch, there would be very few data points with more than one observation. Thus, the frequency distribution would be a long ordered listing of the data points. There is an option on the Frequencies dialog box called "Display frequency tables" that governs whether or not the frequency distribution is displayed. The default for this option is "yes," but you may manually turn off the option by clicking on the box to the right of the phrase.

3.2.2 Quartiles, Deciles, Percentiles and Other Quantiles

In order to obtain quartiles or any percentile, you must use SPSS to create a frequency distribution for the variable and use the cumulative percent column to find the appropriate percentile. For instance, the first quartile (25th percentile) for the "weight" variable in the dieter example shown in Figure 3.3 is 146 pounds, because 145 pounds has a cumulative percent of 24, and 146 pounds a cumulative percent of 28.

(SPSS has procedures that compute quartiles and other percentiles, but at the time of this writing, they do not always result in correct values. The procedure discussed above will always yield an accurate result.)

3.3 The Mean

There are several methods for obtaining the mean of a distribution with SPSS for Windows. You can, of course, use the Frequencies procedure by clicking on **Mean** in the Central Tendency box in the Frequencies: Statistics dialog box. Try this with the dieter data. The mean should be 151.96 pounds. Note that this is very close to the median.

The mean can also be calculated with SPSS using the Descriptives or the Explore procedures. The computer executes Descriptives faster than Frequencies, but it is not possible to obtain the mode or median using the Descriptives procedure. The Explore procedure does not calculate the median, either.

To obtain the mean using the Descriptives procedure:

(1) Click on **Statistics** from the menu bar.

(2) Click on **Summarize** from the pull-down menu.

(3) Click on **Descriptives** from the pull-down menu. This opens the Descriptives dialog box, as shown in Figure 3.4.

Figure 3.4 Descriptives Dialog Box

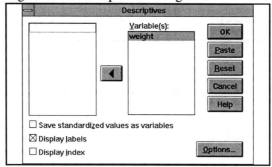

(4) Move the "weight" variable to the Variables box by clicking on the variable and then on the **right arrow button**.

(5) Click on the **Options button** to open the Descriptives: Options dialog box (Figure 3.5).

The default options selected are: Mean, Std. deviation, Minimum and Maximum. If only the mean is required, the other three options may be clicked off as is shown in Figure 3.5.

(6) Click on **Continue** to close this dialog box.

(7) Click on **OK** to run the procedure.

Did you obtain the same mean as you did when you used the Frequencies procedure?

Figure 3.5 Descriptives: Options Dialog Box

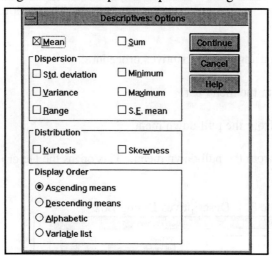

You may also obtain the mean, and several other descriptive statistics, using the Explore procedure as follows:

(1) Click on **Statistics** from the menu bar.

(2) Click on **Summarize** from the pull-down menu.

(3) Click on **Explore** from the pull-down menu.

(4) Click on and move the "weight" variable to the Dependent List box using the **right arrow button**.

(5) In the display box, click on the **Statistics** option.

(6) Click on **OK**.

In addition to the mean, this procedure lists the median and several other descriptive statistics (see Figure 3.6).

Figure 3.6 Output from the Explore Procedure

```
      WEIGHT
Valid cases:         25.0    Missing cases:       .0    Percent missing:          .0

Mean           151.9600    Std Err       1.7516    Min       138.0000    Skewness      .6398
Median         150.0000    Variance     76.7067    Max       172.0000    S E Skew      .4637
5% Trim        151.6444    Std Dev       8.7582    Range      34.0000    Kurtosis     -.2219
95% CI for Mean (148.3448, 155.5752)    IQR        13.0000    S E Kurt      .9017
```

Proportion as a Mean

With a dichotomous variable (a variable with only two possible values, 0 and 1), the mean is the same as the proportion of the cases with a value of 1. We will illustrate this using the "occup.sav" data file. This file contains codes representing the different occupations for 20 primary householders (Table 2.1 in the textbook). Suppose that we wish to find the proportion of respondents who hold professional occupations.

First, we shall create a dichotomy from the "occup" variable, which currently has four levels. To do so, recode all occupations except for professional to a value of 0, representing "non-professional occupations." Leave professional coded 1. (Instructions for recoding variables are given in Section 1.4.2.) Next, compute the mean using the Frequencies procedure, and also obtain the frequency distribution. From the frequency distribution (Figure 3.7), note that the percentage of professionals (coded 1) in the sample is 30%. The mean is given as 0.30, which is just another way of representing 30%.

Figure 3.7 Frequency Distribution and Mean of a Dichotomous Variable

```
OCCUP      occupation

                                                    Valid      Cum
Value Label              Value  Frequency  Percent  Percent   Percent

                           0       14        70.0    70.0      70.0
professional               1        6        30.0    30.0     100.0
                                  -------  -------  -------
                         Total     20       100.0   100.0

Mean          .300

Valid cases    20    Missing cases       0
```

Chapter Exercises

3.1 Table 2.17 of the textbook contains information on the ages at death of English rulers. These data are contained in the "ruler.sav" data file. Use this file to perform the following analyses with SPSS:

(a) Determine the mean, median, and mode of the distribution using the Frequencies procedure. How do these measures compare? What do you think accounts for the differences? Which is closest to what you would call the average age of death today?
(b) What is the 10th percentile? the 90th percentile?
(c) Create a histogram and frequency distribution of the variable. Describe the distribution. Do any of the observations seem to be outliers? If so, how do these observations affect your findings in parts (a) and (b)?

3.2 Use the "football.sav" data file (Table 2.15 in the textbook), which contains heights and weights of 56 football players, to do the following:

(a) Determine the mean height of the sample of football players using either the Frequencies or the Descriptives procedure.
(b) Suppose you discovered that all the players were measured with their shoes on. In order to obtain more accurate heights, subtract 2 inches from each players height, and calculate the mean of the revised heights. (Hint: use the Compute procedure.)
(c) How does your result in (a) compare to that in (b)? What principle does this illustrate?
(d) Repeat the procedure, this time multiplying each player's height by 2. How does the mean of the revised heights compare to the mean of the original heights?

3.3 Figure 5.1 of the textbook contains information on language and non-language IQ scores for a sample of children. Using these data, contained in the file "IQ.sav," use SPSS to complete the following:

(a) Compute the mean, median, and mode of each type of IQ using the Frequencies procedure.
(b) Which is the "best" measure of central tendency for the language IQ scores? Why? For the non-language scores? Why?

3.4 Table 3.6 of the textbook gives a frequency distribution of hypothetical data on income for 64 families. The data are contained in the file "income.sav." The data file contains one variable, "income," with values equal to the midpoints of the income intervals in Table 3.6. That is, 4 families have an income between $0 and $3,999. The midpoint of this range is

$2,000, and this value appears four times in the "income.sav" file.

(a) Compute the median income. In order to accurately compute the median of data displayed in this grouped manner, you must click on the **Values are group midpoints** option in the Frequencies: Statistics dialog box (see Figure 3.1).
(b) Use SPSS to compute the median again, this time clicking off the **Values are group midpoints** option. How do the medians compare?
(c) Does using the midpoint option affect the calculation of the mean of the data? Why or why not?

3.5 The "dieter.sav" data file contains weights of 25 dieters. Following the steps below, use SPSS to illustrate the principle that the sum of all deviations from the mean is zero.

(a) In Section 3.3 we found that the mean weight of this sample of dieters is 151.96 pounds. Compute a new variable, called "devwt," representing deviations from this mean. Details for computing variables are contained in Chapter 1. The algebraic expression, which you will enter in the Numeric Expression box of the Compute Variable dialog box, is: "weight-151.96."
(b) Now, compute the sum and mean of this new variable, "devwt," using the Descriptives procedure.
(c) Did you find that both the sum and mean are 0?

Appendix III
SPSS Syntax for Measures of Location

3.1 The Mode

Use the following command to create a frequency distribution and compute the mode of the variable called "major:"

Frequencies variables=major
/statistics=mode.

3.2.1 The Median

Use the syntax below creates a frequency distribution and compute the median of the variable called "weight:"

Frequencies variables=weight
/statistics=median.

3.3 The Mean

Use either of the commands below to compute the mean of the "weight" variable:

Frequencies variables=weight
/statistics=mean.
           or

Descriptives variables=weight
/statistics=mean.

Proportion as a Mean

Use the following commands to dichotomizes the "occup" variable (recoding old values of 2, 3, and 4 to 0) and then create a frequency distribution and compute the mean for the recoded variable.

Recode occup (2,3,4=0).
Frequencies variables=occup
/statistics=mean.

Chapter 4   MEASURES OF VARIABILITY

In this chapter, we demonstrate how to calculate the range, interquartile range, mean deviation, standard deviation, and standard scores using SPSS. A graphical display of measures of variability is also demonstrated through box-and-whisker plots.

4.1 Ranges

4.1.1 The Range

The range, the difference between the maximum value in a distribution and the minimum value, can be obtained with SPSS for Windows using two different procedures. An easy method is to create a frequency distribution of the scores as demonstrated in Section 2.2.1. Using the Frequencies procedure, look at the top of the distribution to find the minimum value, and then scan down to the bottom value to find the maximum. When you have a small data set with relatively few values, this method is adequate. However, if your data set has many different values, the frequency distribution will be very large and it is more efficient to compute the range using the Statistics option within the Frequencies procedure. First follow steps 1-5 given in Section 2.2 for creating a frequency distribution. Next:

(1) Click on **Statistics** to open the Frequencies: Statistics dialog box (see Figure 2.1).

(2) Click on **Range** in the Dispersion box.

(3) Click on **Continue**.

(4) Click on **OK**.

A second method for obtaining the range and other measures of dispersion uses the Descriptives procedure. This is the preferred approach when you have many different values in the data set. We shall illustrate by finding the range of the federal and state gasoline taxes in Table 2.18 of the text. Begin by opening the "gas.sav" SPSS data file. Next:

(1) Click on **Statistics** on the menu bar.

(2) Click on **Summarize** from the pull-down menu.

(3) Click on **Descriptives** to open the Descriptives dialog box (see Chapter 3, Figure 3.4).

(4) Click on the variable name for which you wish to have the range ("gastax"), and then click on the **right arrow button** to move the variable into the Variables box.

(5) Click on **Options** to open the Descriptives: Options dialog box (Figure 3.5).

(6) In the Dispersion box, click on **Range**. (Notice that the mean, standard deviation, minimum, and maximum boxes are already checked. This is the default.)

(7) Click on **Continue**.

(8) Click on **OK**.

The summary statistics for the Descriptives procedure should appear as shown in Figure 4.1.

Figure 4.1 Descriptives for Gastax Variable

```
Number of valid observations (listwise) =        50.00

                                                      Valid
Variable       Mean      Std Dev    Minimum    Maximum     N   Label

GASTAX        16.96        4.59       4.00      25.00     50
```

The variable name appears on the top left of the output, followed by the descriptive statistics. The range, the difference between the maximum gas tax (25¢) and the minimum (4¢), is 21¢.

4.1.2 The Interquartile Range

The interquartile range, the difference between the first and third quartiles, is less affected by extreme scores than is the range. Because SPSS does not always calculate quartiles correctly (see Section 3.2.2), you should not rely on it to compute the interquartile range. You must compute it by hand, using the frequency distribution to determine the first and third quartiles, and then subtracting to obtain the interquartile range.

4.2 The Mean Deviation

The mean deviation is the average absolute value of the deviations of scores from the mean. SPSS for Windows does not compute the mean deviation directly, but this can be accomplished using alternative procedures.

As an example, we compute the mean deviation using the baby weight data in Table 4.2 in the text. Create a new data file with the five weights, naming the variable "weight" (see

Section 1.3.2). The first step is to calculate the average weight. Do this with the Descriptives procedure as outlined in steps 1-4 in Section 4.1. By default, Descriptives will give you the mean for the weight variable, which in this example is equal to 7 pounds. Once you know the mean, you can calculate the absolute value of the deviations using the Compute procedure.

(1) Click on **Transform** on the menu bar, and then click on **Compute** from the pull-down menu.

(2) Type in the name of the new variable you wish to compute (name this "dev").

(3) Click on **ABS(numexpr)** to calculate the absolute value of the expression.

(4) Click on the "weight" variable and then the **right arrow button**.

(5) Click on the minus sign on the calculator pad, and then click on 7 (the mean).

(6) Click on **OK**.

This will create a new variable in your data file called "dev," obtained as the absolute value of the difference, "weight" minus the mean.

To compute the mean deviation, you now need to calculate the mean of the "dev" variable. This is accomplished with the Descriptives procedure again (steps 1-4, Section 4.1). Click on the "dev" variable, and the mean will automatically appear in the output. The mean should be 1.2, which is the average of the absolute deviations 1, 1, 0, 2, 2.

4.3 The Standard Deviation

The standard deviation is easily calculated by SPSS using the Descriptives procedure.

(1) Click on **Statistics** on the menu bar.

(2) Click on **Summarize** from the pull-down menu.

(3) Click on **Descriptives** to open the Descriptives dialog box (see Figure 3.5).

(4) Click on the variable name that you wish to examine.

By default, the output will contain the mean, standard deviation, minimum, and maximum values. For example, Figure 4.1 contains the default summary statistics for the variable "gastax."

To obtain the variance, the square of the standard deviation, follow steps 1-4 above for the standard deviation, and then:

(1) Click on **Options**.

(2) Click on **Variance** in the Dispersion box.

(3) Click on **Continue** and then click on **OK**.

Notice in your output that the variance is the square of the standard deviation. For the "gaxtax" data in Figure 4.1, the variance is $4.59^2 = 21.07$.

4.5 Some Uses of Location and Dispersion Measures Together

4.5.1 Standard Scores

Standard scores (z-scores) indicate the relative position of any observation in the sample, that is, the number of standard deviations above or below the mean. For values above the mean, the z-score will be positive; values below the mean have negative z-scores; and a value equal to the mean results in a z-score of 0. The z-scores are calculated using the Descriptives procedure.

We shall illustrate using the data on the number of children in 10 families in Table 4.5 of the textbook. Begin by opening the data file "kids.sav." Then:

(1) Click on **Statistics** on the menu bar.

(2) Click on **Summarize** and then **Descriptives** to open the Descriptives dialog box.

(3) Click on the variable name that you wish to standardize ("num_chld"), and then click on the **right arrow button**.

(4) Click on the **Save standardized values as variables** box.

This will cause SPSS to create a new z-score variable. By default, the new variable is named by prefixing the letter z to the first seven letters of the original variable. For example, "num_chld" becomes "znum_chl." You can examine the z-scores in the data file. Notice that some values are positive and some are negative indicating scores that are above and below the mean, respectively. Note also that most z-scores are between -2.0 and $+2.0$. If you calculate the mean and standard deviation of the standard scores (using the Descriptives procedure), you will find that the mean is 0 and the standard deviation is 1.

4.5.2 Box-and-Whisker Plots

Box-and-whisker plots display the median, interquartile range, and extremes of a distribution. SPSS for Windows creates these plots using the Explore procedure. To illustrate this procedure, first create a new data file by entering the ages of 8 senior citizens listed in Section 3.2 in the textbook (these are 60.5, 61, 62, 64, 65, 66, 67, and 69 years). Then:

(1) Click on **Statistics** on the menu bar.

(2) Click on **Summarize** from the pull-down menu.

(3) Click on **Explore** to open the Explore dialog box.

(4) Click on the variable that you wish to plot, and then click the **uppermost right arrow button**.

(5) Click on **Plots** to open the Explore: Plots dialog box.

(6) Click on **Factor levels together** in the boxplots box (this is the default).

(7) Click on **Continue**.

(8) Click on **OK**.

SPSS places the box-and-whisker plot in a Chart Carousel window. To open this window and examine the plot, double click on the **Chart Carousel** icon in the lower left-hand corner of the screen. The box-and-whisker plot will appear as shown in Figure 4.2.

The median of the distribution is indicated by the horizontal line in the middle of the box; here, it is 64.5. You can also find the minimum and maximum values, indicated by the ends of the whiskers. For instance, the top whisker ends at 69 (the maximum), and the lower whisker ends at 60.5 (the minimum).

Figure 4.2 Box-and-Whisker Plot of Ages of Senior Citizens

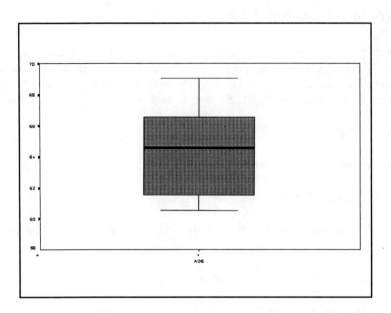

Chapter Exercises

4.1 Using the "head.sav" data file (Table 4.7 of the text) on length and breadth head measurements, perform the following analyses using SPSS for Windows:

(a) Find the range of length of heads using both the Frequencies and Descriptives procedures.
(b) What is the interquartile range of both head measurements?
(c) Compute the standard deviation and variance of head breadth.

4.2 Using the "dieter.sav" data file (Table 2.10 of the text), compute z-scores for the "weight" variable using SPSS.

(a) What is the z-score for the dieter who lost the most weight? The least weight?
(b) Verify that the mean of the z-scores is 0.

4.3 Using SPSS and the data in the "football.sav" file to do the following:

(a) Find the standard deviation of the football players' weights.
(b) Suppose that each player gained 20 pounds. Compute the standard deviation of the new weights (Hint: use the Compute command). Did the standard deviation change? Why or why not?
(c) Suppose that each player went on a radical diet and now weighed one-half of his original weight. Compute the new weights and find the standard deviation of the new weights. How did dividing by a constant affect the standard deviation?

4.4 Use the "phone.sav" data set containing information on telephone calls received by airline reservations offices from 1985 through 1987 (Table 4.8 of the textbook) and SPSS to do the following:

(a) Create a box-and-whisker plot for each of the three years.
(b) What is the median number of calls for each year?
(c) What is the minimum and maximum number of calls for 1987?
(d) Is the 1985 distribution skewed? How can you tell from the graph?
(e) Compute the standard deviation of number of calls for each year. Which year had the least variability in number of telephone calls?

Appendix IV
SPSS Syntax for Measures of Variability

4.1.1 The Range Using Frequencies

To find the range of "gastax" using the Frequencies command, use the following syntax:

Frequencies variables = gastax
 /statistics = range.

To find the range of "gastax" using the Descriptives command, enter the following command:

Descriptives variables = gastax
 /statistics = range.

4.2 The Mean Deviation

The following sequence of commands will calculate the mean deviation for the variable "weight":

Descriptives variables = weight.
Compute dev = ABS(weight − 7).
Descriptives variables = dev.

4.3 The Standard Deviation and Variance

To calculate the standard deviation and variance for any variable, use the following command:

Descriptives variables = varname
 /statistics = stddev variance.

4.5.1 Standard Scores

The following command creates a z-score ("znum_chl") for the variable "num_chld":

Descriptives variables = num_chld (znum_chl).

4.5.1 Box-and-Whisker Plots

The SPSS command for creating a box-and-whisker plot for the variable "age" is:

Examine variables = age
 /plot = boxplot.

Chapter 5 SUMMARIZING MULTIVARIATE DATA: ASSOCIATION BETWEEN NUMERICAL VARIABLES

This chapter describes how to obtain scatter plots and correlation coefficients between numerical variables using SPSS for Windows.

5.1 Association of Two Numerical Variables

5.1.1 Scatter Plots

To illustrate the use of SPSS for scatter plots, let us recreate the plot in Figure 5.1 of the textbook. The data for 23 observations on language and non-language IQ scores are contained in the file "IQ.sav."

To obtain a scatter plot, do the following:

(1) Click on **Graphs** from the menu bar.

(2) Click on **Scatter** from the pull-down menu.

(3) This opens the scatter plot dialog box, which offers four different types of plots[1]. We want the simple plot, which is the default. Therefore, click on **Define**. This opens the Simple Scatterplot dialog box, shown in Figure 5.1.

(4) Click on the variable called "nonlang" and move it to the Y Axis box by clicking on the **uppermost right arrow button**.

(5) Click on the variable called "lang" and move it to the X Axis box by clicking on the **right arrow button**.

(6) Click on **OK** to close this dialog box and create the scatter plot.

---

[1] The "Matrix" option is used for displaying plots of several pairs of variables in one matrix. The "Overlay" option is used for displaying several plots on the same axis. The "3-D" option is used for plotting three variables.

Figure 5.1 Simple Scatterplot Dialog Box

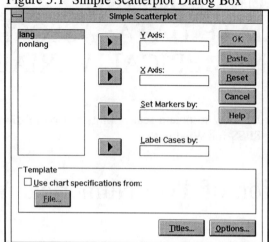

The Chart Carousel window opens automatically since there is no other information in the Output window. Your graph should resemble the one in Figure 5.1 of the textbook, and Figure 5.2 below. There is clearly a positive association between language IQ and non-language IQ; individuals with high language IQ scores also tend to have high non-language IQ scores, and those with low scores on one measure tend to have low scores on the other.

Figure 5.2 Scatter Plot Showing Positive Association

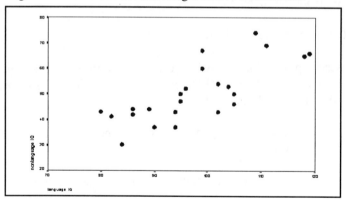

We will create another scatter plot using the data in "lunch.sav." This is a fictitious data set that contains information on the number of calories consumed at lunch by an individual, and the average temperature of the restaurant during the individual's visit. Let us create a scatter plot with temperature on the x-axis and calories on the y-axis. Following steps 1-6 will produce the results shown in Figure 5.3. In this plot, there appears to be almost no association between these two variables. People consume a small, moderate, and large number of calories at lunch regardless of whether the temperature of the restaurant is in low, moderate, or high ranges.

Figure 5.3 Scatter Plot Showing No Association

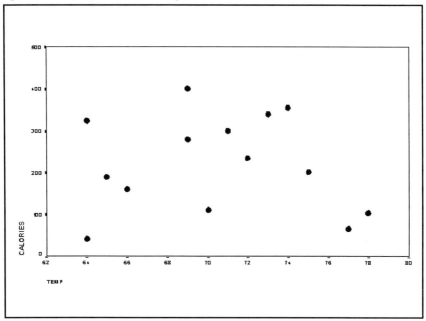

When several observations have the same (x, y) values, the scatter plot can be modified to show these as "sunflowers." This is illustrated in Section 16.1.1 of this Guide.

Changing the Scales of the Axes

SPSS choses the scales for the x and y axes that best fit the range of the data, but you may manually adjust the scales if you wish. Below are steps to edit the x-axis scale of any scatter plot:

(1) In the Chart Carousel window, click on **Edit** to open the Chart window.

(2) Click on **Chart** from the menu bar.

(3) Click on **Axis** from the pull-down menu to open the Axis Selection dialog box.

(4) Select the X Scale and click on **OK** to open the X Scale Axis dialog box (Figure 5.4).

Figure 5.4 X Scale Axis Dialog Box

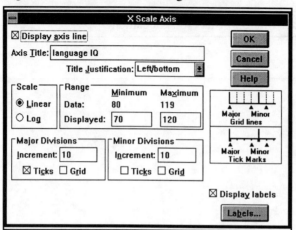

(5) Edit the Minimum and Maximum parts of the Range box as desired and click on **OK** to close the dialog box and re-draw the scatter plot.

5.1.2 Other Information Revealed by Scatter Plots

Examining a scatter plot can reveal information that may not be readily apparent from the correlation coefficient. You can discern, for instance, when a relationship between two variables is nonlinear and you can locate bivariate outliers. We will illustrate that latter of these cases.

The data file "IQ.sav" contains information regarding language and non-language IQ scores for 23 students. The range of the language scores is 80 to 119, and the range of the non-language scores is 30 to 74. The scatter plot of these two variables shows a positive association.

Now, open the data file "IQ2.sav." (See Section 1.3.1 for details on opening files.) This file contains the same scores in the original "IQ.sav" file, plus one additional data point, a student with a language score of 80 and a non-language score of 72. Considering each of these scores alone, neither is an outlier; each is within the range of the original scores for its variable.

Create the scatter plot of these scores (see Figure 5.5) and locate the point (80,72). It represents an individual with a very low language IQ and a very high non-language IQ. There are no other data points in its vicinity, and it is clearly an outlier. The data analyst should attempt to understand why this unusual pairing of values occurred.

Figure 5.5 Scatter Plot Showing a Bivariate Outlier

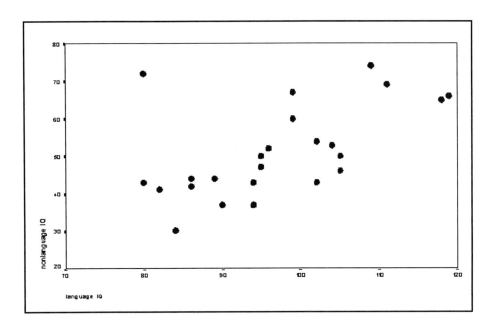

5.1.3 The Correlation Coefficient

The correlation coefficient summarizes the degree of linear association between two numerical variables. The procedure to calculate the correlation coefficient with SPSS is straightforward; we will illustrate using the "conform.sav" data file. This file contains information regarding conformity ratings of husbands and wives, and is taken from Table 5.10 in the textbook. You may begin by opening the file.

To compute the correlation coefficient for these two measured variables:

(1) Click on **Statistics** from the menu bar.

(2) Click on **Correlate** from the pull-down menu.

(3) Click on **Bivariate** from the pull-down menu. This opens the Bivariate Correlations dialog box (see Figure 5.6).

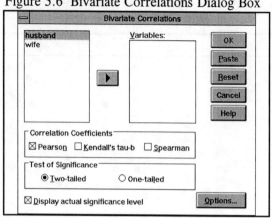

Figure 5.6 Bivariate Correlations Dialog Box

(4) Click on the "husband" and "wife" variables and move them to the Variables box by clicking on the **right arrow button**.

(5) For the moment, we are not concerned with the significance level of the correlation, so click off the **Display actual significance level** box.

(6) All of the other defaults are acceptable, so click on **OK**.

The output should look like that in Figure 5.7.

Figure 5.7 Correlation of Husband and Wife Conformity Ratings

```
                    - - Correlation Coefficients  - -

              HUSBAND      WIFE

HUSBAND       1.0000      .8491**
WIFE           .8491**   1.0000

* - Signif. LE .05     ** - Signif. LE .01      (2-tailed)

" . " is printed if a coefficient cannot be computed
```

SPSS lists the correlation coefficients it calculates in a correlation matrix. The values on the diagonal are all 1.0 because they represent the correlations of each variable with itself. The values above and below the diagonal are identical. In the example, the correlation between husbands' and wives' conformity ratings is .8491. This is a strong, positive association; married couples tend to have similar degrees of conformity.

Let us repeat this procedure two more times, first with the "IQ.sav" data file, and then with the "IQ2.sav" data file. For the original data file, you should determine that the correlation is .7689. Simply by adding one outlying observation ("IQ2.sav"), the correlation coefficient decreases to .5611. This is a dramatic change, and underscores the need to examine the scatter plot of two variables in addition to calculating the coefficient.

5.1.4 Rank Correlation

When data are ordinal instead of being recorded on an interval or ratio scale, it is appropriate to calculate the Spearman correlation coefficient for ranks. Using the data from Table 5.3 of the textbook, we will use SPSS to rank order the exposure and mortality variables and then compute the correlation coefficient. When data are gathered that are already on an ordinal scale (e.g., 1=best, 2=second best, ..., 10=worst), the first step in the example ("ranking") is not necessary.

Open the "cancer.sav" data file. The two variables in this file are "expose," the index of exposure to radioactive materials for the county, and "mortalit," the rate of cancer mortality per 100,000 person-years.

To rank these two variables:

(1) Click on **Transform** from the menu bar.

(2) Click on **Rank Cases** from the pull-down menu. This opens the Rank Cases dialog box (see Figure 5.8).

Figure 5.8 Rank Cases Dialog Box

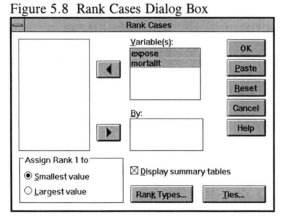

(3) Move the "expose" and "mortalit" variables to the Variable(s) box by clicking on the name

of each variable and then on the **right arrow button**.

(4) Click on **OK**.

SPSS creates two new variables, "rexpose" and "rmortali" which consist of ranked data. (SPSS automatically creates the new variable names by adding an r-prefix to the original variable names.) The procedure to compute the Spearman correlation for these two variables is similar to the steps outlined for calculating the Pearson coefficient. The only difference is that you must click off the Pearson option and click on the Spearman option in the Correlation Coefficients box (see Figure 5.6).

Your output should look like the following:

Figure 5.9 Spearman Correlation Listing

```
From       New
variable   variable   Label
--------   --------   -----

EXPOSE     REXPOSE    RANK of EXPOSE
MORTALIT   RMORTALI   RANK of MORTALIT




 - - -   S P E A R M A N   C O R R E L A T I O N     C O E F F I C I E N T S   - - -


RMORTALI            .8333**

             REXPOSE


*  - Signif. LE .05     ** - Signif. LE .01     (2-tailed)

" . " is printed if a coefficient cannot be computed
```

The correlation of mortality and exposure, when ranked, is .8333, which is the same as that calculated in the textbook. Note that the output also provides information regarding the transformation procedure.

5.2 More than Two Variables

Correlation Matrix

SPSS displays the Pearson correlation between two variables in a correlation matrix. This is also the case when you obtain correlations between all pairs of variables when there are more than two of them.

The procedure is the same as that detailed in Section 5.1.3, but you need to include the names of all of the variables for which you desire correlations in the Variables box of the Bivariate Correlations dialog box.

To illustrate, open the "fire.sav" data file and create a correlation matrix for the following variables: "agility," "body," "composit," "obstacle," "stair," and "written" by including the names of all of them in the Variables box. (To obtain the same results as those listed below you should, click off the box labeled "Display Actual Significance Levels.") Your output should look like Figure 5.10.

Figure 5.10 Correlation Matrix

```
                - - Correlation Coefficients - -

           AGILITY      BODY     OBSTACLE     STAIR      WRITTEN     COMPOSIT

AGILITY    1.0000     .9617**     .8752**    .9538**     -.4578*     -.9702**
BODY        .9617**   1.0000      .7591**    .9062**     -.4660*     -.9401**
OBSTACLE    .8752**    .7591**   1.0000      .7343**     -.4951**    -.8748**
STAIR       .9538**    .9062**    .7343**   1.0000       -.3374      -.8983**
WRITTEN    -.4578*    -.4660*    -.4951**   -.3374       1.0000       .6596**
COMPOSIT   -.9702**   -.9401**   -.8748**   -.8983**      .6596**    1.0000

 * - Signif. LE .05      ** - Signif. LE .01       (2-tailed)

 " . " is printed if a coefficient cannot be computed
```

Although this matrix is large, it is not difficult to read. For instance, the correlation between the body drag test and the obstacle course is at the intersection of the column labeled body and the row labeled obstacle, and also at the intersection of the column labeled obstacle and the row labeled body. The correlation between these two variables (.7591) is shown in boldface type in Figure 5.9.

It is also possible to discern patterns among correlations in the matrix. For example, the correlations among body drag, stair climb, and obstacle course times are all positive and moderately strong (ranging from .7343 for obstacle course with stair climb, to .9062 for stair climb with body drag). Because all are measures of athletic behavior, the positive association is to be expected. The written test score is negatively correlated with these tasks, that is, high scores on the written test are associated with low times on the agility tasks. Thus, more agile applicants tend to have higher scores on the written test, and vice versa.

Missing Values

Only those cases with values for both variables can be used in computing a correlation coefficient. There are two ways to cause SPSS to eliminate cases with missing values: "listwise deletion" and "pairwise deletion." (See Chapter 1 for a discussion of missing values.) As an example, suppose that the third case in our data file were missing a score on the written test. Because the third case does not have complete information on all of the variables, listwise deletion eliminates the third case from the computation of all correlation coefficients. All correlations are calculated from the remaining 27 cases. Pairwise deletion eliminates the third case when computing only those correlations that involve the written test; thus some coefficients would be based on 27 observations and others would be based on all 28.

The default option in SPSS is pairwise deletion which uses the maximum number of cases for each coefficient. You may request listwise deletion be clicking on the **Options button** of the Bivariate Correlations dialog box (Figure 5.6).

Chapter Exercises

5.1 Table 4.7 in the textbook gives data on the head measurements (length and breadth) for 25 infants. The data are contained in the "head.sav" data file. Use this file and SPSS to do the following:

(a) Make a scatter plot of the length and breadth variables. Which data point or points, if any, would you consider to be outliers?
(b) Determine the correlation coefficient between the two variables. Comment on the nature of the relationship, including the strength and direction of association.
(c) If you noted any outliers in part (a), how do you think this may have affected your results in part (b)?

5.2 It has been shown that the relationship between amount of stress and work productivity is curvilinear. In other words, extremely low and extremely high amounts of stress are related to low work productivity, but moderate amounts of stress are associated with the maximum amount of productivity.

(a) Create a hypothetical data set with two variables -- amount of stress and work productivity -- that you think will illustrate this relationship. Your data file should have a minimum of 20 observations.
(b) Use SPSS to produce a scatter plot of your data. Did you succeed in simulating a curvilinear relationship?
(c) Name three other instances in which you might find a curvilinear relationship between two variables.

5.3 Using the "conform.sav" data file:

(a) Use SPSS to rank the "husband" and "wife" variables, and to compute the Spearman correlation coefficient.
(b) Comment on the direction and magnitude of the relationship, and how it compares to the Pearson correlation computed in Section 5.1.3.

5.4. Using the "lunch.sav" data file, perform the following analyses using SPSS:

(a) Suppose that each of the individuals in the sample was offered (and accepted) a free slice of chocolate cake with his or her lunch. Therefore, you must add 300 calories to each persons calorie count. (Hint: use the Compute procedure.) Find the correlation coefficient between the "revised" calorie count and temperature of the restaurant. How does it compare to the correlation coefficient of the original variables?
(b) Now suppose that a similar study was conducted in Buffalo, NY, in January. Coincidentally, the individuals sampled in Buffalo ate the same number of calories as did those from the first sample. The only difference was that the temperature of each of the restaurants was exactly ½ that of the first sample. Create a new variable representing the temperature of the Buffalo restaurants.
(c) Compute the correlation of Buffalo temperatures with caloric intake. Does the correlation differ from the original? Why or why not?
(d) Suppose that the people in Buffalo were also offered the slice of cake. What do you think the correlation of the two "new" variables will be, that is, the correlation between the increased caloric intake and the reduced temperature variables? Verify your prediction by using SPSS for Windows to compute the correlation coefficient.

5.5 The data file "enroll.sav" contains information from a random sample of 26 school districts. Information was obtained on the following variables:

(1) district enrollment;
(2) the percentage of students in the district who are African-American;
(3) the percentage of students who pay full price for lunches;
(4) an index of racial disproportion in classes for emotionally disturbed children (which is positive if the proportion of African-American students is greater than the proportion

of white students).

Using this data file, compute a correlation matrix for all four variables and use it to answer the following questions:

(a) Which correlation is largest (in magnitude)?
(b) Explain what is meant by the negative correlation between enrollment and percent of African-Americans.
(c) Racial disproportion is most highly associated with which other variable? What is the magnitude and direction of the association?

Appendix V
SPSS Syntax for Summarizing Multivariate Data: Association Between Numerical Variables

5.1.1 Scatter Plots

Use the following syntax to create a scatter plot of the variables "lang" (on the x-axis) and "nonlang" (on the y-axis):

Plot plot nonlang with lang.

5.1.3 The Correlation Coefficient

Use the following command to compute the Pearson correlation coefficient between the variables "husband" and "wife:"

Correlations variables=husband wife.

5.1.4 Rank Correlation

The following command computes the Spearman rank correlation between the variables "rexpose" and "rmortali." Although it is possible to direct SPSS to rank measured variables, the process using syntax is long and complex. Therefore, this command assumes that the variables were entered into the data file as ranks.

Nonpar corr variables=rexpose rmortali.

Correlation Matrix

Use the following syntax to create a correlation matrix of the variables listed:

Correlations variables=agility body obstacle stair written composit.

Chapter 6 SUMMARIZING MULTIVARIATE DATA: ASSOCIATION BETWEEN CATEGORICAL VARIABLES

In this chapter, we demonstrate how to examine associations between two or among more than two categorical variables. These methods rely on frequency tables, which represent the counts and patterns of association.

6.1 Two-by-Two Frequency Tables

The summarization of data in a frequency table is called cross-tabulation because the information is categorized using two classifications simultaneously. For example, we may have a sample with certain numbers of women and men in it. We may also have numbers of whites and minorities in the sample. Recall from Chapter 2 that we could find the frequencies for one variable using the Frequencies procedure. However, it may not be immediately apparent how many white females or minority males there are in the data file. This is accomplished using the Crosstabulation procedure, which examines the counts of simultaneous occurrences of several values.

For example, open the data in the firefighter data file ("fire.sav") which has performance information on firefighter candidates. To create a two-way frequency table of race by sex, follow these steps:

(1) Click on **Statistics** from the menu bar.

(2) Click on **Summarize** from the pull-down menu.

(3) Click on **Crosstabs** to open the Crosstabs dialog box shown in Figure 6.1.

(4) Highlight the "sex" variable by clicking on it, and then move it to the Row(s) box by clicking on the **right arrow button**.

(5) Highlight the "race" variable by clicking on it, and then move it to the Column(s) box by clicking on the **right arrow button**.

(6) Click on **OK**.

By default, this will produce a crosstabulation of the row ("sex") by column ("race") variables. The number of cases for each combination of values of the row by column variables is displayed in the cells of the table. Your output should look like Figure 6.2.

Figure 6.1 Crosstabs Dialog Box

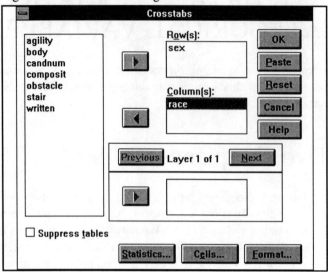

Figure 6.2 Crosstabulation of Sex by Race

```
SEX    by    RACE

                      RACE               Page 1 of 1
              Count  |
                     | white      minority
                     |                         Row
                     |     1  |       2  |   Total
SEX                  -------------------------
                 1   |     9  |       5  |     14
    male             |        |          |    50.0
                     -------------------------
                 2   |     8  |       6  |     14
    female           |        |          |    50.0
                     -------------------------
              Column      17         11         28
              Total     60.7       39.3      100.0

Number of Missing Observations:   0
```

By examining the frequency table, we see that sex has two values (male and female) and race has two values (white and minority). Race is the column variable and sex is the row variable. This two-by-two table has four "cells."

The number that appears in each of the cells is called the Count (or frequency); this is the number of cases in that cell. In this sample, there are 8 white females and there are 5 minority males. The numbers to the right of the table and below the table are the marginal totals and represent the counts for the rows and columns separately. For example, in the row margins, the top 14 indicates that there are 14 males and the bottom 14 indicates that there are 14 females. Likewise, the 17 and 11 column totals indicate the number of whites and minorities, respectively. The 28 in the bottom right corner indicates the total number of cases in the sample. These marginal frequencies could also be obtained using the Frequencies procedure as shown in Section 2.2.

Calculation of Percentages

In addition to the count for each cell, SPSS will also calculate row, column, and total percentages. The row percentage is the percentage of cases in a row that fall into a particular cell. The column percentage is the percent of cases in a column that fall into a particular cell. The total percentage is the number of cases in a cell expressed as a percentage of the total number of cases in the table.

To calculate row, column, and total percentages, follow steps 1-5 in the preceding section, and then:

(1) Click on **Cells** to open the Crosstabs: Cell Display dialog box.

(2) Click on each box in the Percentages box to indicate that you wish you calculate **Row**, **Column**, and **Total** percentages.

(3) Click on **Continue**.

(4) Click on **OK**.

To illustrate these percentages, enter the data on 25 students that appear in Table 6.1 of the textbook into a new data file. Code the categorical data as follows: sex (1=female; 2=male), division (1=graduate; 2=undergraduate). Run the Crosstabs procedure with "sex" as the row variable and "division" as the column variable. After calculating row, column, and total percentages, your output should appear as in Figure 6.3.

Figure 6.3 Crosstabulation of Sex by Division with Percentages

```
SEX   by   DIVISION

                        DIVISION       Page 1 of 1
            Count     |
            Row Pct   |grad       undergr
            Col Pct   |                         Row
            Tot Pct   |    1.00|      2.00| Total
SEX         ------------------------------
               1.00   |       8 |       4 |    12
  female              |    66.7 |    33.3 |  48.0
                      |    72.7 |    28.6 |
                      |    32.0 |    16.0 |
                      ------------------------------
               2.00   |       3 |      10 |    13
  male                |    23.1 |    76.9 |  52.0
                      |    27.3 |    71.4 |
                      |    12.0 |    40.0 |
                      ------------------------------
            Column          11        14       25
            Total         44.0      56.0    100.0

Number of Missing Observations:  0
```

The cell counts indicate that there are 8 female graduate students, 4 female undergraduates, 3 male graduate students, and 10 male undergraduates. The marginal counts indicate that there are 12 females and 13 males, and that there are 11 graduate students and 14 undergraduates. The total number of cases in the table is 25 which appears in the bottom right hand corner of the table.

The marginal percentages represent the percentage of the total that is found in the row or column. For example, of the entire sample of 25 students, 12 of them (48%) are females. Also, 56% of all of the students are undergraduates.

The top number in each cell is the count. The second number is the row percentage (Row Pct), which represents the percentage of the row that is found in the cell. For example, 23.1% of all males are graduate students (100 x 3/13), and 33.3% of all females are undergraduates (100 x 4/12). The third number in each cell is the column percentage (Col Pct) which is the percentage of the column total in the cell. For example, 72.7% of the graduate students are females (100 x 8/11). Also, 71.4% of the undergraduates are males (100 x 10/14). The fourth line of each cell contains the total percentage (Tot Pct). For example, 3 male graduate students represent 12% of all 25 students in the table. This was computed by dividing 25 by 3.

Phi-Coefficient

The phi-coefficient is an index of association for a two-by-two table. To compute the phi-coefficient, follow steps 1-5 in Section 6.1, and then:

(1) Click on **Statistics** to open the Crosstabs: Statistics dialog box.

(2) Click on **Phi and Cramer's V** in the Nominal Data box.

(3) Click on **Continue**.

(4) Click on **OK**.

For example, the value of the phi-coefficient for the data in Figure 6.3 is .439.

6.2 Larger Two-Way Frequency Tables

In many studies, categorical variables have more than two values. The number of categories is not limited to two, and virtually any size row-x-column table is possible. For example, "religion" may be recorded in three categories: Catholic, Jewish, and Protestant; "occupation" may have 15 categories representing 15 different job titles. The procedure for calculating counts and percentages, as well as the interpretations of the frequencies, is the same as described in Section 6.1.

As an example, open the "spit.sav" data file. This is a study on the effectiveness of two interventions to help major league baseball players stop using spit tobacco.[1] The frequency table for this example is a 3 x 2 table; there are two variables, "outcome" with three levels and "intervention" with two levels.

---

[1] Data reproduced from summary tables in Greene, J.C., Walsh, M.M., & Masouredis, C. (1994). Report of a pilot study: A program to help major league baseball players quit using spit tobacco. *Journal of the American Dental Association, 125,* 559-567.

Figure 6.4 Crosstabulation of Outcome by Intervention

```
OUTCOME   outcome of intervention   by   INTERVEN   intervention group

                         INTERVEN         Page 1 of 1
             Count    |
             Row Pct  | minimum    extended
             Col Pct  |                          Row
             Tot Pct  |   1.00|       2.00|    Total
OUTCOME               ---------------------------
               1.00   |           |        5  |      5
   quit                |           |    100.0  |    9.3
                      |           |     19.2  |
                      |           |      9.3  |
                      ---------------------------
               2.00   |        3  |       12  |     15
   tried              |     20.0  |     80.0  |   27.8
                      |     10.7  |     46.2  |
                      |      5.6  |     22.2  |
                      ---------------------------
               3.00   |       25  |        9  |     34
   failed             |     73.5  |     26.5  |   63.0
                      |     89.3  |     34.6  |
                      |     46.3  |     16.7  |
                      ---------------------------
             Column          28          26         54
             Total         51.9        48.1      100.0

Number of Missing Observations:   0
```

Using this data set, follow the steps in Section 6.1 to create a frequency table containing counts, and row, column, and total percentages. Your results should look like Figure 6.4. The row variable is player outcome (quit; tried to quit; failed to quit), and the column variable is the length of intervention (minimum or extended). The number of players who quit successfully is 5. There are 34 players who failed to quit, and 73.5% of them are in the minimum intervention group. Overall, would you say that the players were successful in quitting? Probably not, since only 9.3% had quit successfully. There are approximately equal numbers of players in both intervention groups (28 and 26). It appears as though the minimum length intervention was not as effective as the extended length since there were no minimum intervention players who quit successfully, and 73.5% of the players who failed to quit were in the minimum intervention group.

6.3 Three Categorical Variables

6.3.1 Organization of Data for Three Yes-No Variables

We will now examine relationships among three categorical variables. The data for this example come from a study of crowding and antisocial behavior in 75 community areas in Chicago (see Table 6.45 of the textbook). Three characteristics of the communities are cross-classified to examine the relationships among socioeconomic status ("SES"), population density ("pop_dens"), and delinquency rate ("delinq"). Each variable is dichotomous, and has been coded as 1=low and 2=high. Thus, this is a 2 x 2 x 2 table (i.e., three variables, each with 2 levels). This data file is named "delinq.sav."

Table 6.1 Crosstabulation of 75 Communities by Delinquency, Population Density, and Socioeconomic Status

	Low SES		High SES	
	Low population density	High population density	Low population density	High population density
Low delinquency	3	2	27	3
High delinquency	2	33	2	3

The 2 x 2 x 2 frequency table is reproduced in Table 6.1. This is a three-way table displayed as a pair of 2 x 2 tables, one for the high-SES communities and one for the low-SES communities.

The approach we will demonstrate for understanding the associations among the variables is to examine the "marginal" association between the pairs of variables. The marginal association involves examining 2 x 2 frequency tables for three pairs of variables: SES x pop_dens, SES x delinq, and pop_dens x delinq. The three tables appear in Figure 6.5. In order to have SPSS for Windows produce these three tables, you need to create three separate 2 x 2 tables using the steps in Section 6.1.

Figure 6.5 Separate Crosstabulations for Three Categorical Variables

POP_DENS population density by DELINQ rate of juvenile delinquency

```
                       DELINQ         Page 1 of 1
              Count  |
              Row Pct| low      high
              Col Pct|                    Row
              Tot Pct|    1.00|    2.00|  Total
POP_DENS      --------------------------
                1.00 |     30  |      4 |    34
    low        |     88.2     11.8       45.3
               |     85.7     10.0
               |     40.0      5.3
              --------------------------
                2.00 |      5  |     36 |    41
    high       |     12.2     87.8       54.7
               |     14.3     90.0
               |      6.7     48.0
              --------------------------
              Column        35       40      75
              Total       46.7     53.3   100.0

Number of Missing Observations:  0
```

SES by POP_DENS population density

```
                       POP_DENS       Page 1 of 1
              Count  |
              Row Pct| low      high
              Col Pct|                    Row
              Tot Pct|    1.00|    2.00|  Total
SES           --------------------------
                1.00 |      5  |     35 |    40
    low        |     12.5     87.5       53.3
               |     14.7     85.4
               |      6.7     46.7
              --------------------------
                2.00 |     29  |      6 |    35
    high       |     82.9     17.1       46.7
               |     85.3     14.6
               |     38.7      8.0
              --------------------------
              Column        34       41      75
              Total       45.3     54.7   100.0

Number of Missing Observations:  0
```

(Continued)

SES   by   DELINQ   rate of juvenile delinquency

```
                         DELINQ          Page 1 of 1
                Count    |
                Row Pct  |  low       high
                Col Pct  |                        Row
                Tot Pct  |   1.00|     2.00|    Total
SES             ---------------------------------
                 1.00    |     5  |      35 |     40
   low                   |  12.5  |    87.5 |   53.3
                         |  14.3  |    87.5 |
                         |   6.7  |    46.7 |
                         ---------------------------
                 2.00    |    30  |       5 |     35
   high                  |  85.7  |    14.3 |   46.7
                         |  85.7  |    12.5 |
                         |  40.0  |     6.7 |
                         ---------------------------
                Column        35          40        75
                 Total       46.7        53.3     100.0
```

Number of Missing Observations: 0

The top table shows the relationship between delinquency and population density. The second and third tables portray the relationship of socioeconomic status with population density and delinquency, respectively. To illustrate, the data in the middle table show the following: of the 75 communities in the sample, 34 (45.3%) have low population density. Of these 34 low-population communities, only 5 (14.7%) have lower socioeconomic status conditions.

Table 6.1 gives more detail than any of the 2 x 2 marginal crosstabulations. For example, among high SES communities, 30 (27 + 3) have low levels of delinquency, of which 27 (90%) are in low population density areas. SPSS will not produce a three-way table directly, but does so by producing separate 2 x 2 tables for each value of the third variable (e.g., high- and low-SES communities, separately). This "conditional" approach to examining frequency tables is discussed in Sections 6.3.2 and 6.4.

6.3.2 Larger Three-Way Frequency Tables

Figure 6.6 is a 3 x 3 x 2 table containing hypothetical data on restaurant characteristics. Restaurants were classified according to type of meal they served (pasta, French, seafood), cost of the meal (inexpensive, moderate, expensive), and whether or not the restaurant was part of a chain (yes, no). The data are contained in "meal.sav."

The interrelationships are displayed in two 3 x 3 frequency tables, one for chain and one for non-chain restaurants. These tables are produced with SPSS for Windows using the following procedures. After opening the data file:

(1) Click on **Statistics** from the menu bar.

(2) Click on **Summarize** from the pull-down menu.

(3) Click on **Crosstabs** to open the Crosstabs dialog box.

(4) Highlight the "meal" variable by clicking on it, and then move it to the Row(s) box by clicking on the **top right arrow button**.

(5) Highlight the "cost" variable by clicking on it, and then move it to the Column(s) box by clicking on the **middle right arrow button**.

(6) Highlight the "chain" variable by clicking on it, and then move it to the blank box near the bottom of the screen with the **bottom right arrow button**.

(7) Click on **OK**.

Figure 6.6 Crosstabulation of Meal by Cost for Each Value of Chain

```
MEAL    type of meal    by    COST    cost of meal
Controlling for..
CHAIN    chain restaurant    Value = 1.00    yes

                            COST                              Page 1 of 1
                 Count    |
                 Row Pct  | inexpens  moderate  expensiv
                 Col Pct  | ive                  e            Row
                 Tot Pct  |    1.00|     2.00|     3.00| Total
MEAL             ---------------------------------------------
                   1.00   |     5  |     1  |     2  |    8
  pasta                   |  62.5  |  12.5  |  25.0  |  36.4
                          |  55.6  |  20.0  |  25.0  |
                          |  22.7  |   4.5  |   9.1  |
                          -------------------------------
                   2.00   |     1  |     2  |     4  |    7
  french                  |  14.3  |  28.6  |  57.1  |  31.8
                          |  11.1  |  40.0  |  50.0  |
                          |   4.5  |   9.1  |  18.2  |
                          -------------------------------
                   3.00   |     3  |     2  |     2  |    7
  seafood                 |  42.9  |  28.6  |  28.6  |  31.8
                          |  33.3  |  40.0  |  25.0  |
                          |  13.6  |   9.1  |   9.1  |
                          -------------------------------
                 Column         9        5        8       22
                 Total       40.9     22.7     36.4    100.0        (Continued)
```

```
MEAL   type of meal   by  COST   cost of meal
Controlling for..
CHAIN   chain restaurant   Value = 2.00   no
                          COST                         Page 1 of 1
              Count    |
              Row Pct  |inexpens moderate expensiv
              Col Pct  |ive               e              Row
              Tot Pct  |     1.00|     2.00|     3.00|  Total
MEAL          ---------+---------+---------+---------+
                 1.00  |    4    |    2    |    1    |    7
  pasta               |  57.1   |  28.6   |  14.3   |  29.2
                      |  66.7   |  28.6   |   9.1   |
                      |  16.7   |   8.3   |   4.2   |
              ---------+---------+---------+---------+
                 2.00  |    1    |    2    |    7    |   10
  french              |  10.0   |  20.0   |  70.0   |  41.7
                      |  16.7   |  28.6   |  63.6   |
                      |   4.2   |   8.3   |  29.2   |
              ---------+---------+---------+---------+
                 3.00  |    1    |    3    |    3    |    7
  seafood             |  14.3   |  42.9   |  42.9   |  29.2
                      |  16.7   |  42.9   |  27.3   |
                      |   4.2   |  12.5   |  12.5   |
              ---------+---------+---------+---------+
              Column        6         7        11        24
              Total       25.0      29.2      45.8     100.0

Number of Missing Observations:   0
```

Each table displays the relationship between type and cost of meal for one type of restaurant. (Notice that the first table is for chain restaurants and the second table is for non-chain restaurants.) The marginal counts and percentages reveal that chain restaurants had approximately the same numbers of pasta, french, and seafood restaurants. The most common cost categories were inexpensive and expensive. Among non-chain restaurants, there are more French restaurants than either pasta or seafood, and more expensive restaurants (45.8%) than inexpensive (25%) or moderate (29.2%).

6.4 Effects of a Third Variable

The association between two variables may be examined at specific values of a third variable, that is, the "conditional" association between two variables. The relationship between two variables may be maintained, increased, decreased, or even reversed when a third variable is taken into account.

As an example, let's again look at the data on juvenile delinquency rates and population density; the frequencies in Table 6.2 are obtained by summing the low-SES and high-SES counts in Table 6.1. This figure is the same as the first crosstabulation in Figure 6.5 except that the columns and rows are reversed.

Table 6.2 Crosstabulation of Delinquency by Population

		POP_DENS		
		Low	High	All
DELINQ	Low	30	5	35
	High	4	36	40
	All	34	41	

In Table 6.2 we see that most of the low-delinquency communities are located in low-density areas, while most of the high-delinquency communities are in high-density areas. Note also that the numbers of high and low delinquency types are quite similar (35 and 40) as are the numbers of high and low population (34 and 41).

Now, let's examine the relationship between delinquency and population when the third variable is taken into account; this variable is socioeconomic status (SES). Using SPSS for Windows, separate tables for low-SES and high-SES communities are obtained by using the following steps.

(1) Click on **Statistics** from the menu bar.

(2) Click on **Summarize** from the pull-down menu.

(3) Click on **Crosstabs** to open the Crosstabs dialog box.

(4) Highlight the "delinq" variable by clicking on it, and then move it to the Row(s) box by clicking on the **top right arrow button**.

(5) Highlight the "pop_dens" variable by clicking on it, and then move it to the Column(s) box by clicking on the **middle right arrow button**.

(6) Highlight the "SES" variable by clicking on it, and then move it to the blank box near the bottom of the screen with the **bottom right arrow button**.

(7) Click on **OK**.

The results are displayed in Figure 6.7. Note that these tables have the same results as given in the lefthand and righthand portions of Table 6.1, respectively.

The patterns in Figure 6.7 are different from those in Table 6.2. First, notice that the row and column counts have changed considerably when SES is considered. Among low SES areas, there are many more high than low delinquency communities, and most of these (94.3%) are in high-density areas. Among high-SES communities, there is a greater percentage of low delinquency (85.7%) than high delinquency, and most of these are in low-density areas. In a sense, SES "explains" the association between population density and juvenile delinquency rates.

Figure 6.7 Crosstabulation of Delinquency by Population for Each Level of SES

```
DELINQ   rate of juvenile delinquency  by  POP_DENS  population
density
Controlling for..
SES   Value = 1.00   low

                       POP_DENS        Page 1 of 1
                Count  |
                Row Pct| low       high
                Col Pct|                        Row
                Tot Pct|    1.00|     2.00| Total
DELINQ          --------------------------
                1.00   |     3   |     2   |     5
   low          |    60.0 |    40.0 |  12.5
                |    60.0 |     5.7 |
                |     7.5 |     5.0 |
                --------------------------
                2.00   |     2   |    33   |    35
  high          |     5.7 |    94.3 |  87.5
                |    40.0 |    94.3 |
                |     5.0 |    82.5 |
                --------------------------
        Column          5         35        40
         Total       12.5       87.5     100.0
```

(Continued)

```
DELINQ   rate of juvenile delinquency   by   POP_DENS   population
density
Controlling for..
SES   Value = 2.00   high

                         POP_DENS          Page 1 of 1
              Count   |
              Row Pct |   low       high
              Col Pct |                          Row
              Tot Pct |    1.00|     2.00|    Total
DELINQ                ----------+---------+
              1.00    |     27  |      3  |      30
   low                |    90.0 |    10.0 |    85.7
                      |    93.1 |    50.0 |
                      |    77.1 |     8.6 |
                      ----------+---------+
              2.00    |      2  |      3  |       5
   high               |    40.0 |    60.0 |    14.3
                      |     6.9 |    50.0 |
                      |     5.7 |     8.6 |
                      ----------+---------+
              Column         29         6         35
              Total        82.9      17.1      100.0
```

Chapter Exercises

6.1 Read in the data in "semester.sav" and use SPSS to create a crosstabulation of students' major by the number of statistics courses taken. Obtain the cell counts, and row, column, and total percentages.

(a) What percentage of the students were biology majors?
(b) What percentage of biology majors took three semesters of statistics?
(c) How many students took only one semester of statistics?
(d) What percentage of students had no statistics?
(e) Which major took the greatest number of statistics courses?

6.2 Using the "fire.sav" data, use SPSS to do a crosstabulation of race and sex and answer the following questions:

(a) What percentage of all firefighters were minority females?
(b) Were there more minority or white male firefighters?
(c) Calculate the phi-coefficient. How strong is the relationship between race and

gender?

6.3 Using the "meal.sav" data file, create crosstabulations with SPSS to do the following:

(a) Examine the association between type of food ("meal") and expense ("cost") for each category of the "chain" variable (i.e., conditional association). (Create separate 3 x 3 tables for chain and non-chain restaurants.)

(b) What percentage of the restaurants are not seafood restaurants? Is this the same for both chain and non-chain restaurants?

6.4 Using SPSS for Windows, create a two-way frequency table of population density ("pop_dens) by delinquency ("delinq") using the "delinq.sav" data file.

(a) Does the relationship change once you control for SES?
(b) Discuss the effects of the third variable ("SES").

Appendix VI
SPSS Syntax for Summarizing Multivariate Data: Association Between Categorical Variables

6.1 Two-by-Two Frequency Tables

   To create a 2 x 2 frequency table with just counts, enter the following command:

   Crosstabs tables = race by sex.

 Calculation of Percentages and Phi Coefficient

   Use the following syntax for a 2 x 2 frequency table with row, column, and total percentages, and the phi coefficient:

   Crosstabs tables = sex by division
   /cells = count row column total
   /statistics=phi.

6.2 Larger Two-Way Frequency Tables

   For larger two-way tables, use following command:

   Crosstabs tables = interven by outcome
   /cells=count row column total.

6.3 Three Categorical Variables

   To create three two-way tables, use the following three commands:

   Crosstabs tables = pop_dens by delinq
   /cells=count row column total.
   Crosstabs tables = ses by pop_dens
   /cells=count row column total.
   Crosstabs tables = ses by delinq
   /cells=count row column total.

6.4 Effects of a Third Variable

   Use the following single command to examine the effects of a third variable (the relationship between delinquency and population density, controlling for ses).

   Crosstabs tables = delinq by pop_dens by ses
   /cells=count row column total.

PART III

PROBABILITY

Chapter 7 BASIC IDEAS OF PROBABILITY

Although SPSS for Windows is designed primarily to be used for data analysis and not for evaluating probability functions per se, it is possible to demonstrate certain probability concepts with the program. This chapter illustrates "tossing coins" and "rolling dice" using SPSS. All of the procedures discussed in this chapter involve sampling with replacement.

7.3 Probability in Terms of Equally Likely Cases

The Bernoulli distribution, discussed in Chapter 9 of the textbook, is the distribution of a variable that can take one of two values -- 0 or 1. There are several Bernoulli distributions, differing with respect to the probability associated with the values. If the probability is .5 that a random draw from this distribution will be a 1 (and .5 that it will be a 0), we can say that sampling from this distribution is the same as tossing a fair coin. Here, obtaining a value of 1 corresponds to tossing a "head," and a value of 0 corresponds to a "tail" (or vice versa).

To simulate coin tossing on SPSS:

(1) Once in SPSS, click on **File** from the menu bar.

(2) Click on **New** from the pull-down menu.

(3) Click on **Data** from the pull-down menu.

(4) You need to compute a variable that has a value of 1 or 0 (a Bernoulli variable). SPSS will not permit you to compute any variable without having an active data set, however. To create such a data set, you must type some number (e.g., 1) in the first cell of the first column of the data file.

(5) Click on **Transform** from the menu bar.

(6) Click on **Compute** from the pull-down menu to open the Compute Variable dialog box (see Figure 7.1).

Figure 7.1 Compute Variable Dialog Box

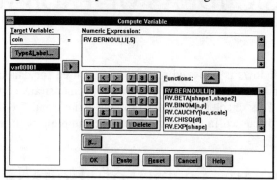

(7) Type in the name of the new variable (e.g., "coin") in the Target Variable box.

(8) Locate the Functions box and scroll down through the options until you find "RV.BERNOULLI(p)." Highlight this function and move it to the Numeric Expression box by clicking on the **up arrow button**.

(9) You will notice that the "p" turns into a question mark, prompting you to select a specific Bernoulli distribution by indicating the probability of obtaining a value of 1. Type .5 where the question mark is.

(10) Click on **OK** to run the procedure.

Note that the value of the new variable, "coin," has a value of 1 or 0. This represents the result of the coin toss.

You can also simulate rolling a die with a similar procedure. The random variable that corresponds to this is from a discrete uniform distribution, with values ranging from 1 to 6. SPSS has only the continuous uniform distribution as a random variable function. Therefore, to make SPSS roll a die, follow steps 1-6 above, and then:

(1) Type in the name of the new variable (e.g., "die") in the Target Variable box.

(2) Click on "RND(numexpr)" in the Functions box and move it to the Numeric Expression box by clicking on the **up arrow button**. This is the round function, which rounds continuous decimal numbers to integers.

(3) The (numexpr) will change to a (?), prompting you to indicate the value or expression you wish to round. Locate and highlight the "RV.UNIFORM(min,max)" in the Function box, and move it to the Numeric Expression box by clicking on the **up arrow button**. It should automatically replace the question mark from the RND expression.

(4) You will notice that the min and max have now changed to question marks. Type 1,6 in this space, and delete any extra commas or question marks.

(5) Click on **OK** to run this procedure.

In the column labeled die, there will be an integer between 1 and 6, representing the roll of the die.

7.8 Random Sampling; Random Numbers

Random Numbers

You can repeat the procedure in Section 7.3 several times to generate several random numbers from a distribution. To do so, simply modify step (4) in Section 7.3 by entering a number for multiple cases for the variable. For instance, if you want to perform 10 coin tosses, you need to enter some number, all 1's for instance, in the first 10 cells of the first column. Then, compute the "coin" variable exactly as described above. You will obtain 10 values, each either 0 or 1, representing the result of each of ten coin tosses.

Chapter Exercises

7.1 Following the instructions in Section 7.8, simulate 20 coin tosses with SPSS.

(a) Did you obtain 10 heads (1's) and 10 tails (0's)? If not, why not?
(b) Would you have come closer to an equal split with 50 tosses? with 10 tosses? Why or why not?

7.2 Think of a number between 1 and 5. Direct SPSS to pick a number at random within the range 1 through 5. (Hint: use the uniform distribution).

(a) Did the two numbers coincide? What is the probability that they would be the same?
(b) Repeat this procedure 10 times. How many matches were there? How many did you expect to obtain (based on the law of probability)?

Appendix VII
SPSS Syntax for Basic Ideas of Probability

7.3 Probability in Terms of Equally Likely Cases

The following commands simulate rolling a die. The expression containing the uniform function determines a continuous random number from a uniform distribution ranging from 0 to the number in parentheses. The expression containing the rnd function modifies the values to be integers from 1 to 6, representing the die face.

Compute unif = uniform(5).
Compute die = (rnd(unif) + 1).
Execute.

Chapter 8 PROBABILITY DISTRIBUTIONS

In this chapter, we demonstrate how to use SPSS for Windows to generate probability distributions. We will concentrate on exploring properties of the standard normal distribution, but SPSS can perform the same tasks for other probability distributions as well.

8.5 Family of Standard Normal Distributions

Finding Probability for a Given z Value

When examining probability distributions, it is important to be able to determine the percentage of the distribution that lies within a certain interval. We know, for instance, that 50% of the area under the standard normal curve lies below the point 0. In other words, the probability that a random variable drawn from a standard normal distribution will be less than 0 is .50.

Similarly, referring to Figure 8.7 in the textbook, we see that 15.9% (13.6% + 2.2% + 0.1%) of the area under the standard normal curve lies below −1. Again, the probability of obtaining a number less than -1 on a random draw from this probability distribution is .159.

The cumulative distribution function in SPSS for Windows will compute these probabilities. The procedure for obtaining these computations is detailed below:

(1) Click on **File** from the menu bar.

(2) Click on **New** from the pull-down menu.

(3) Click on **Data** from the pull-down menu.

(4) SPSS will not permit you to compute a new variable without having an active data set. So, "activate" the Data window by typing some number (e.g., 1) in the first cell of the first column of the data file.

(5) Click on **Transform** from the menu bar.

(6) Click on **Compute** from the pull-down menu.

(7) Type in the name of the new variable (e.g., "probilty") in the Target Variable box.

(8) Highlight the "CDF.NORMAL(q,mean,stddev)" function in the Functions box and move it into the Numeric Expression box by clicking on the **up arrow button**.

(9) The parameters in the parentheses will be replaced by question marks. The first question mark (the "q" parameter) represents the point on the distribution for which you wish to obtain a cumulative probability estimate. For the first example, we will find the cumulative probability distribution for 0. Therefore, modify the expression to read "CDF.NORMAL(0,0,1)."

(10) Click on **OK** to run the procedure.

SPSS will return the value of .5 as the value of the "probilty" variable.

Repeat this procedure, this time for the value -1. You can do so by changing the "q value" from 0 to -1. (Note: When you click on **OK** SPSS will prompt you to indicate whether or not you wish to "Change Existing Variable?" Click on **OK** to run the procedure.) The value of the "probilty" variable should change to .16.

Finding a z Value for a Given Probability

There may also be instances when you wish to determine the point on a probability distribution associated with a given cumulative probability. For instance, the value of 0 has 50% of the standard normal distribution below it. The function in SPSS that will return the value of 0 for input of .5 is the inverse distribution function.

The procedure for obtaining this result is similar to that outlined previously. Follow steps 1-6 above, and then:

(1) Type in the name of the new variable (e.g., "value") in the Target Variable box.

(2) Highlight the "IDF.NORMAL(prob,mean,stddev)" function and move it to the Numeric Expressions box by clicking on the **up arrow button**.

(3) To find the value on the standard normal distribution that has 95% of the values below it, enter the value .95 for the probability parameter, the value of 0 for the mean parameter, and the value of 1 for the stddev parameter.

(4) Click on **OK** to run the procedure.

You should obtain the value of 1.64.

Chapter Exercises

8.1 Using SPSS, find the value on the standard normal distribution that has 2.3% of the distribution below it. Does it correspond to the information given in Figure 8.7 of your textbook?

8.2 Sketch the curve of a normal distribution with mean 2 and standard deviation 1. How much of this distribution is below the value 2? Verify your answer using SPSS.

8.3 Estimate the value on the standard normal distribution that has 60% of the distribution below it. Use SPSS to evaluate your estimate.

8.4 Repeat Exercise 8.2 using a normal distribution with mean 0 and standard deviation 2.

8.5 Using SPSS, determine the proportion of the standard normal curve that is:

(a) above -1.
(b) between -2 and 2.

Hint: SPSS will not compute these areas directly; you must perform some simple computations by hand.

Appendix VIII
SPSS Syntax for Probability Distributions

8.5 Finding Probability for a Given z Value

Use the following to compute a variable, called "probilty," that represents the probability that a random variable from the standard normal distribution is less than or equal to 1:

Compute probilty=cdfnorm(1).
Execute.

8.5 Finding a z Value for a Given Probability

Use the following command will compute a variable, called "value," that represents the value of the standard normal distribution that has 50% of the distribution below it:

Compute value=probit(.5).
Execute.

Chapter 9 SAMPLING DISTRIBUTIONS

In this chapter, we use SPSS to simulate drawing random samples from probability distributions. We also examine some of the properties of the sampling distributions of the sum and the mean, and the Central Limit Theorem.

9.1 Sampling from a Population

Random Samples

Although it may not have occurred to you, when we make SPSS toss a coin 10 times (Section 7.8), we are actually taking a random sample from a Bernoulli probability distribution. Now we shall repeat this procedure, but with a sample of 50 items from a standard normal distribution. Because the procedure is so similar to those described in Chapters 7 and 8, the following instructions are abbreviated.

(1) Open a new Data window (see Section 7.3).

(2) Enter a "1" for the first 50 cases of the first column.

(3) Click on **Transform** from the menu bar.

(4) Click on **Compute** from the pull-down menu.

(5) In the Compute dialog box, name the Target Variable "sample."

(6) Highlight the "RV.NORMAL(mean,stddev)" function in the Functions box and move it into the Numeric Expression box by clicking on the **up arrow button**.

(7) Choose a standard normal distribution by replacing the first and second question marks with 0 and 1, respectively.

(8) Click on **OK**.

Now create a histogram of the "sample" variable (see Section 2.3.1 for details on creating histograms). It should resemble a standard normal distribution, but will differ somewhat because it is a sample. Figure 9.1 shows one possible histogram. This histogram would appear more normal if we had taken a larger number of samples or created the histogram with a smaller number of intervals. Because your random sample is different, your histogram will vary somewhat. Note, for instance, that the mean for this sample is .11, which is close to the mean of 0 and the standard deviation of this random sample (.90) is close to 1.

Figure 9.1 Histogram from a Standard Normal Distribution

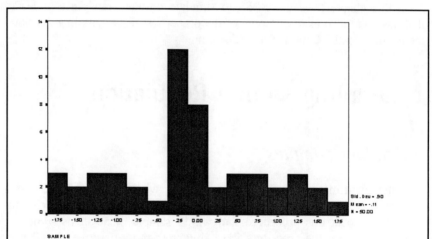

9.2 Sampling Distribution of a Sum and of a Mean

In this section we obtain the sampling distribution of the sum of two variables from a discrete uniform distribution with minimum of 1 and maximum of 6. This is analogous to constructing the sampling distribution resulting from rolling two dice.

We will direct SPSS to roll these dice, one at a time, for a total of 50 pairs of rolls. Next, we will compute the sum of each roll (e.g., the first roll for die one + the first roll for die two, etc.) and then examine the frequency distribution of this new variable.

(1) Open a new Data window, and type a "1" in the first 50 rows of the first column.

(2) Click on **Transform** from the menu bar.

(3) Click on **Compute** from the pull-down menu.

(4) Name the target variable "sample1."

(5) In the Numeric Expression box, create the expression "RND(RV.UNIFORM(1,6))."

(6) Compute another variable in the same manner. Label it "sample2."

(7) Compute a "total" variable which is the sum of the two sample variables. Do so by clicking on the "sample1" variable and moving it to the Numeric Expression box with the **right arrow button**, clicking on the + from the calculator pad, and then moving the "sample2" variable to the Numeric Expression box with the **right arrow button**.

(8) Obtain a histogram of the "total" variable (see Section 2.3.1). It should somewhat resemble the histogram in Figure 9.2 of your textbook.

You can also obtain the sampling distribution of the mean of two (or more) variables in a similar manner.

9.5 The Normal Distribution of Sample Means

The Central Limit Theorem

It is also possible to use SPSS to illustrate the Central Limit Theorem (CLT). The process is not straightforward, but because the CLT is one of the most important principles of inferential statistics, working through this example may help you to understand the concepts more fully.

The central limit theorem asserts that the distribution of the sample mean of a certain number of observations will resemble a normal distribution. This is true regardless of the parent distribution from which the samples are drawn.

To illustrate this theorem, we will first have to obtain a random sample of size n (e.g., 50) from a specific distribution (e.g., discrete uniform(1,10)), then repeat this process many times (e.g., 100), calculate the mean for each of the random samples, and finally inspect the frequency distribution and histogram of the sample means.

The procedures used to obtain a random sample from a specific distribution are given in Chapter 7. The tedious part of this process involves repeating the sampling 100 times. We have completed this step for you, and the results are saved in the data file "clt.sav." Retrieve this file. You will note that the first variable is entitled "marker." This is simply a place holder used to obtain the desired sample size of 50. There are 100 more variables, u1 to u100, which represent the 100 times that SPSS took a random sample of size 50. Therefore, at present we have a [50 x 100] matrix representing [sample size x number of samples].

In order to get a histogram of the means, we first need to calculate the mean of each of the 100 samples. We could direct SPSS to compute the mean of each of the "u" variables separately, but we would then have to manually input each of these means into another column. A less time consuming method is to transform the matrix in such a way as to make SPSS keep track of the means. To do so, we have to transpose the matrix -- interchange the rows and the columns -- and then compute the means. Open the "clt.sav" data file, and then:

(1) Click on **Data** from the menu bar.

(2) Click on **Transpose** from the pull-down menu.

(3) Highlight all of the variable names except "marker" (click on **u1**, hold the mouse button down, and drag down to the name of the last variable in the list).

(4) Move the variable names to the Variable(s) box by clicking on the **upper right arrow button**.

(5) Click on **OK**.

(6) You will receive a message indicating that you have not included all the variables in the transpose command, and that all variables not included will be lost. This is referring to the "marker" variable. Click on **OK**.

You should now have a transposed data file with a [100 x 50] matrix (excluding the first column). The rows now represent the 100 samples drawn, and the columns represent the 50 random variables selected in each sample.

We can now compute the mean of each of the rows. To do this:

(1) Click on **Transform** from the menu bar.

(2) Click on **Compute** from the pull-down menu.

(3) Enter "mean" in the Target Variable box.

(4) In the Numeric Expression box, create the expression: "mean(var001 to var050)." (The easiest way to do this is to type it in the box.)

(5) Click on **OK**.

This should create a new variable, "mean," which contains the means of the 100 samples. Create a histogram of this new variable. Figure 9.2 displays this graph. Although the distribution of the means is not exactly normal, it is very roughly normal. Normality would be improved if the parent distribution were more normal-like or if we had drawn more than 100 samples.

Figure 9.2 Histogram of Means

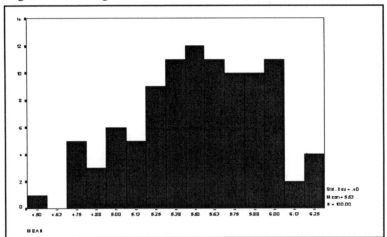

Chapter Exercises

9.1 Use SPSS to:

(a) Obtain 10 random samples of size 8 from a standard normal distribution, and then compute the mean of each of these samples. Are all or any of the sample means equal to 0, the mean of the population? Sketch a histogram of these means and summarize the distribution.
(b) Repeat part (a), using 10 random samples of size 40.
(c) Compare your results in part (a) and part (b). What principle do these results illustrate?

9.2 Use SPSS to:

(a) Obtain 3 random samples of size 10 from a Bernoulli distribution with $p=.5$.
(b) Compute a new variable which represents the sum of these three original variables.
(c) Based on probabilities, sketch the histogram you would expect to obtain for this composite variable.
(d) Create the histogram with SPSS and compare the actual results to those in part (c). How would you explain the differences?

PART IV

STATISTICAL INFERENCE

Chapter 10 USING A SAMPLE TO ESTIMATE CHARACTERISTICS OF ONE POPULATION

This chapter demonstrates how to use SPSS to estimate characteristics of a population. Both point estimates and interval estimates are discussed, although SPSS does not complete the computations for many different confidence intervals. Remember that statistical inference involves estimating *population* characteristics on the basis of the information from a random *sample*; that is, SPSS bases all of its calculations on sample values contained in the data set.

10.1 Estimation of a Mean by a Single Number

The sample mean is a point estimate of the population mean. The standard error of the mean is a measure of how precise the estimate is.

The sample mean and standard error can be obtained using three different procedures. You can use the Descriptives procedure (Section 3.3), the Frequencies procedure (Section 3.3), or the Explore procedure (Section 3.3). By default, all three procedures compute the mean, although the standard error is the default only with the Explore procedure. To compute the standard error with the Frequencies and Descriptives procedures, additional steps are necessary as outlined below.

To use the Frequencies procedure to compute the mean and standard error:

(1) Click on **Statistics** from the menu bar.

(2) Click on **Summarize** from the pull-down menu.

(3) Click on **Frequencies** to open the Frequencies dialog box.

(4) Click on the name of the variable that you wish to examine.

(5) Click on the **right arrow button** to move the variable name into the Variables box.

(6) Click on **Statistics** to open the Frequencies: Statistics dialog box.

(7) Click on **S.E. mean** in the Dispersion box.

(8) Click on **Continue**.

(9) Click on **OK**.

To use the Descriptives procedure to compute the mean and standard error, follow these steps:

(1) Click on **Statistics** on the menu bar.

(2) Click on **Summarize** from the pull-down menu.

(3) Click on **Descriptives**.

(4) Click on the desired variable and then the **right arrow button** to move it into the variable(s) box.

(5) Click on **Options** to open the Descriptives: Options dialog box.

(6) Click on **S.E. mean** in the Dispersion box.

(7) Click on **Continue**.

(8) Click on **OK**.

Using the Descriptives procedure with the data on percentage of minority enrollment in 30 school districts (Table 2.16 in the textbook), the output should appear as shown in Figure 10.1. (The sample mean, standard deviation, minimum, and maximum are computed by default.) The sample mean is 18.10 and the standard error of the mean, labeled S.E. Mean, is 2.46. Like the mean, the standard error is an estimate computed from sample data.

Figure 10.1 Summary Statistics from the Descriptives Procedure

```
Number of valid observations (listwise) =        30.00
                                                              Valid
Variable        Mean S.E. Mean    Std Dev   Minimum   Maximum   N   Label
ENROLL          18.10      2.46     13.50      4.00     63.00   30
```

10.2 Estimation of Variance and Standard Deviation

It is also possible to estimate the population variance and standard deviation from sample data. As with the mean, the variance and standard deviation are easily estimated using either the Descriptives, Frequencies, or Explore procedures. The Descriptives procedure is demonstrated in Section 4.3. The sample variance and standard deviation are computed by default using the Explore procedure. When using the Frequencies procedure, you need to follow steps 1-6 in Section 10.1, and then:

(1) Click on **Std. deviation** and **Variance** in the Dispersion box.

(2) Click on **Continue**.

(3) Click on **OK**.

By default, the Frequencies procedure automatically prints the frequency distribution. If you want to suppress this so that only the summary statistics are printed, make sure that **Display Frequency Tables** is not clicked "on."

10.3 An Interval of Plausible Values for a Mean

Section 10.1 demonstrates estimation of a population mean by a single number. It is also possible to obtain an interval estimate of the mean. A confidence interval is obtained from the sample mean by adding to it and subtracting from it a certain number of standard errors.

The standard error of the mean is $\sigma/\sqrt{n}$ where σ is the population standard deviation. In some research, the population standard deviation is known and can be used directly in obtaining the confidence interval. In many studies, however, the value of σ is not known and the sample standard deviation (s) may be substituted. Unfortunately, SPSS for Windows always bases its computations on the sample standard deviation. Thus, Section 10.3.1 illustrates the computation of a confidence interval from partial information that can be obtained from SPSS. Section 10.3.2 illustrates how to complete SPSS results for a confidence interval based on the sample standard deviation.

10.3.1 Confidence Intervals when the Standard Deviation is Known

Based upon the normal distribution, we know that the probability is .95 that the sample mean is within 1.96 standard errors of the population mean. Thus, there is a .95 probability of drawing a sample such that the interval from the mean minus 1.96 standard errors (the lower limit) to the mean plus 1.96 standard errors (the upper limit) contains the population mean. When this principle is applied to a single sample, the resulting interval is the confidence interval, and .95 (or 95%) is the confidence coefficient. It is possible to have confidence coefficients other than .95. For example, a 90% interval is obtained from the sample mean by adding and subtracting 1.64 standard errors.

The SPSS program bases its calculations on the sample standard deviation (s). There are some situations, however, when the population standard deviation is known. For example, standardized intelligence tests have standard deviations that are preset by the publisher ($\sigma = 15$); the Scholastic Assessment Tests (SAT) and Graduate Record Exam (GRE) have preset standard deviations in the population ($\sigma = 100$). When a population standard deviation is available, it should be used in confidence intervals (and in hypothesis tests). In this situation, part of the work must be done "by hand."

We illustrate this using the IQ data for 23 children given in "IQ.sav." We are going to calculate a 95% confidence interval for the mean of language IQ, assuming the population standard deviation is known. After opening the data file:

(a) Obtain the mean of language IQ using the Descriptives procedure. In this example, the mean of language IQ is 97.57.

(b) Using a calculator, compute the upper limit by: Upper = $97.57 + 1.96(15/\sqrt{23})$.

(c) Using a calculator, compute the lower limit by: Lower = $97.57 - 1.96(15/\sqrt{23})$.

Your calculations should lead you to conclude with 95% confidence that the population mean is between 91.44 and 103.70.

10.3.2 Confidence Intervals when the Standard Deviation is Estimated

In most situations, we do not have a value for the population standard deviation, and the confidence interval may be computed using the sample standard deviation. This is done easily with SPSS since the program bases its calculations on the sample standard deviation.

As an example, open the data file "fire.sav" to find the confidence interval for the mean of the stair climbing times ("stair"):

(1) Click on **Statistics** from the menu bar.

(2) Click on **Summarize** from the pull-down menu.

(3) Click on **Explore**.

(4) Click on the "stair" variable and move it into the dependent list box with the **top right arrow button**.

(5) Click on **Statistics** to open the Explore: Statistics dialog box.

(6) Click on **Descriptives: Confidence Interval for Mean**. Note that 95% is the default. You may change this if you require a different level of confidence by moving your cursor to this box and typing in the desired level.

(7) Click on **Continue**.

(8) Click on **Statistics** in the display box. This is an optional step, but recommended since it suppresses unnecessary tables in your output.

(9) Click on **OK**.

Your output will appear as shown in Figure 10.2. First note that the sample mean is 16.27. The confidence interval is labeled "95% CI for Mean" with a lower limit value of 14.7977 and an upper limit of 17.7380. We are 95% confident that the population mean is in the range 14.80 to 17.74.

Figure 10.2 Explore Output with 95% Confidence Interval

```
    STAIR

Valid cases:        28.0    Missing cases:      .0   Percent missing:       .0


Mean         16.2679  Std Err       .7165  Min      11.7000  Skewness      1.7626
Median       15.1000  Variance    14.3749  Max      29.1000  S E Skew       .4405
5% Trim      15.8849  Std Dev      3.7914  Range    17.4000  Kurtosis      3.9022
95% CI for Mean (14.7977, 17.7380)        IQR       3.7000  S E Kurt       .8583
```

10.4 Estimation of a Proportion

A proportion can be viewed as the mean of a dichotomous variable scored 1 or 0. Calculating the mean of a dichotomous variable is accomplished in SPSS using the Descriptives procedure. To illustrate, calculate the mean of the "race" variable in the "fire.sav" data file as follows:

(a) Open the data file as described in Section 1.3.1.

(b) Recode the "race" variable so that whites are assigned a value of 1 and minorities a value of 0, using the Recode procedure as described in Section 1.4.2.

(c) Obtain the mean using the Descriptives procedure as described in Section 3.3.

The mean of this recoded variable is now the proportion of whites in the sample, and the point estimate of the proportion in the population. In this example, the mean is equal to 0.61 which means that approximately 61% of the population is white.

SPSS for Windows does not compute an appropriate standard error or confidence interval for a proportion. The 95% interval is given by Eqn 10.2 in the textbook. This must be computed by hand. For the firefighter example, the standard error of the proportion of whites is $\sqrt{(.61 \times .39)/28} = 0.092$.

10.5 Estimation of a Median

10.5.1 Point Estimation of a Median

An estimate of the population median is obtained by finding the median of a random sample from that population. The sample median is readily obtained using either the Frequencies or Explore procedure. Both procedures are demonstrated in Section 3.2.1.

10.5.2 Interval Estimation of a Median

SPSS will not readily construct a confidence interval for the population median. However, you can obtain a confidence interval by using the program to obtain the median and a frequency distribution, and performing several computations by hand. For example, suppose we wish to construct the confidence interval for the median score of 28 obstacle course times in the "fire.sav" data set. After opening the data file:

(a) Arrange the "obstacle" variable in ascending order and find the median using the Frequencies procedure (see Section 2.2).

(b) By hand, calculate r using the approximate formula for a 95% confidence interval: $r = (n + 1)/2 - 0.98\sqrt{n}$. With n = 28, this result is r = 9.3, which rounds to 9.

(c) Now go back to the frequency distribution and locate the value of "obstacle" at the 9th (x_r) position and the 20th (x_{n-r+1}) position. These are the lower and upper limits for your confidence interval. Thus, we are 95% confident that the population median of obstacle is between 99.5 and 124.0.

10.6 Paired Measurements

10.6.1 Mean of a Population of Differences

When members of a single sample are measured twice (as in "pre" and "post" studies) we can compute a difference score, representing the first score compared to the second score. From this, it is possible to obtain point and interval estimates of the population mean difference.

To demonstrate this analysis:

(a) Enter the pupil reading data in Table 10.2 in the textbook into a new data file. Enter the "before" and "after" score for each case.

(b) Use the Compute command to compute the difference score (after - before) as demonstrated in Section 1.4.1.

(c) Follow the steps in Section 10.3.2 to obtain a 95% confidence interval for the mean of the difference variable. Your output should appear as shown in Figure 10.3. The confidence interval is interpreted as showing that, with 95% confidence, the mean improvement in reading in the population is between 0.32 and 0.70 units. Note that the average gain is 0.51 (the point estimate).

Figure 10.3 Confidence Interval for Reading Score Difference

```
        DIFF

Valid cases:         30.0    Missing cases:       .0    Percent missing:        .0


Mean              .5100    Std Err       .0897    Min       -.2000    Skewness       .9546
Median            .3500    Variance      .2416    Max       1.6000    S E Skew       .4269
5% Trim           .4852    Std Dev       .4915    Range     1.8000    Kurtosis       .1666
95% CI for Mean (.3265, .6935)                    IQR        .5250    S E Kurt       .8327
```

10.6.2 Matched Samples

Some paired measurements are based on matched samples such as two members of a dyad. Difference scores then represent a comparison of the two members.

The SPSS procedure for estimating a confidence interval for the difference score is the same as that given in Section 10.6.1, except that the scores are from two members of a matched pair. You may illustrate using the "conform.sav" data file to construct the confidence interval for the mean difference in conformity between husbands and wives. The point estimate of the difference is -2.30. The 95% confidence interval indicates that the mean difference is between -4.27 and -0.33 units. All plausible values for the difference are negative, indicating that wives exhibit greater conformity. In this example, if we used a 90% confidence level, the values are -3.93 and -0.67. The 90% interval is slightly narrower than the 95% interval.

Chapter Exercises

10.1 Using the "phone.sav" data file, use SPSS to:

(a) Find the mean and standard deviation of the number of phone calls for each year (1985, 1986, 1987). Hint: use the Select If procedure.
(b) Are these the means of the population or sample?
(c) Construct a 95% confidence interval for the mean of each year. Interpret the interval for the year 1985.
(d) Which year had the widest confidence interval? Why?

10.2 Using the "dieter.sav" data file, use SPSS to:

(a) Compute 90%, 95%, and 99% confidence intervals for the mean weight of dieters.
(b) Could the company marketing the diet reasonably claim that dieters will lose an average of 10 pounds? 20 pounds? Why, or why not?

10.3 Use the data on crowding and antisocial behavior in 75 Chicago communities in the file "delinq.sav." Recode the "delinquency" and "SES" variables so that 0=low and 1=high, and then use SPSS to:

(a) Calculate the proportion of high-delinquency communities.
(b) Compute a 90% confidence interval for the proportion of high-SES communities in the U.S., assuming that the data file contains results for a random sample of such communities.

Appendix X
SPSS Syntax for Using a Sample to Estimate Characteristics of One Population

10.1 Estimation of a Mean by a Single Number
    To obtain the mean and the standard error or the mean of the variable "enroll," use either one of the following two commands:

    Descriptives variables = enroll
        /statistics=mean semean.

    Frequencies variables = enroll
        /statistics=mean semean.

10.2 Estimation of Variance and Standard Deviation
    To calculate the variance and standard deviation of the variable "enroll," use either one of the following procedures:

    Descriptives variables = enroll
        /statistics=stddev variance.

    Frequencies variables = enroll
        /statistics=stddev variance.

10.3.1 Confidence Intervals when the Standard Deviation is Known
    To calculate the confidence interval when the standard deviation is unknown, first find the mean of the variable ("IQ"), and then use that mean in the equation to calculate the upper and lower bounds. For example:

    Descriptives variables = IQ.
    Compute upper = 97.57 + (1.96*(15/sqrt(23))).
    Compute lower = 97.57 − (1.96*(15/sqrt(23))).
    Execute.

10.4 Estimation of a Proportion
    To get the point estimation of a proportion, first recode the variable ("race") so that the values are 0 and 1, and then calculate the mean of the recoded variable using the following commands:

    Recode race (1=1) (2=0).
    Descriptives variables=race.

10.5 Estimation of a Median
    To find the point estimation of a median, use the following command:

    Frequencies variables= varname
        /statistics=median.

10.6 Paired Measurements
    To calculate the mean of a population of differences, first compute a difference score ("diff") and then find the mean of the difference score:

    Compute diff = (after − before).
    Descriptives variables = diff.

Chapter 11 ANSWERING QUESTIONS ABOUT POPULATION CHARACTERISTICS

This chapter describes how to use SPSS to test hypotheses about characteristics of a single population. SPSS for Windows has a procedure that computes the test statistic for a one-sample test only when the population standard deviation (σ) is unknown. When σ is known, you must compute the test statistic by hand, but you can use SPSS to find the sample mean, proportion, or median. This is especially useful when sample sizes are large and manual calculations are cumbersome.

11.1 Testing a Hypothesis About a Mean

11.1.1 Hypothesis Testing Procedures

SPSS for Windows does not directly conduct a test of a single mean when the population standard deviation (σ) is known. Situations in which σ is known arise when test scores have standard deviations that are preset by the publisher (e.g., intelligence tests, standardized tests of academic achievement, SAT or GRE tests). They also occur when data have been recorded for many thousands of cases, for example, when industrial processes are monitored over a period of years or when medical records have been compiled for thousands of patients. Nevertheless, the SPSS Frequencies procedure (Section 3.3), the Descriptives procedure (Section 3.3), or the Explore procedure (Section 3.3) may be used to obtain the sample mean ($\bar{x}$); the remaining computations must be performed by hand.

To illustrate, we will use the language IQ scores of 23 pupils in the data file "IQ.sav." Suppose that we wish to test the hypothesis $H_0: \mu \geq 100$ against the alternative $H_1: \mu < 100$. Running any of the three procedures yields $\bar{x} = 97.6$. The sample standard deviation printed by SPSS is not used in calculating the test statistic because the population standard deviation of the IQ test has been set to $\sigma = 15$ by the publisher. By hand we compute the test statistic $z = (97.6 - 100) \div 15/\sqrt{23} = -0.77$. The significance point for the one-tailed test at the 5% level, from the standard normal table, is -1.645; thus the null hypothesis is accepted. (If we had been making a two-tailed test, the significance points would be -1.96 and 1.96; in this example the same conclusion would follow.)

11.1.2 Validity Conditions

It is good practice to check the data for normality prior to conducting the test of significance. A visual way to inspect for normality is to plot a histogram of the variable. There is an option that directs SPSS to impose a normal curve on the graph, which makes it easier to evaluate the normality of the data. We will illustrate this using the language score of the "IQ.sav" data file. You can obtain this graph using the Frequencies procedure as follows:

(1) Click on **Statistics** from the menu.

(2) Click on **Summarize** from the pull-down menu.

(3) Click on **Frequencies** from the pull-down menu.

(4) Click on and move the "lang" variable to the variables box of the Frequencies dialog box by clicking on the **right arrow button**.

(5) Click on **Charts** to open the Frequencies: Charts dialog box.

(6) Click on **Histogram** and **With Normal Curve** in the Chart Type box.

(7) Click on **Continue** to close the dialog box.

(8) Click off the option labeled **Display Frequency Table** in the lower lefthand corner of the Frequencies dialog box.

(9) Click on **OK**.

Your output should look like that shown in Figure 11.1.

Figure 11.1 Sample Histogram with Normal Curve Superimposed

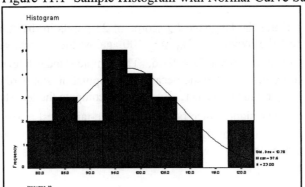

Although the distribution is not precisely normal, it is not highly skewed either. The histogram fits fairly well under the normal curve superimposed on the graph. Because the test is fairly "robust" with respect to normality, we conclude that it is an appropriate method for hypothesis testing in this application.

11.3 Testing Hypotheses About a Mean when the Standard Deviation is Unknown

When the standard deviation of the population is not known, you will need to estimate σ in order to compute the test statistic. The SPSS program computes the sample standard deviation (s) as well as the appropriate test statistic (t). The procedure for using SPSS is the same regardless of whether you are making a one-tailed or a two-tailed test.

We will illustrate using the "football.sav" data file, which contains heights and weights of 56 Stanford football players. Treating this year's team as a sample from a population of football players, we will test the hypothesis that the mean height of football players is equal to 6 feet (72 inches). The null and alternative hypotheses are $H_0: \mu = 72$ and $H_1: \mu \neq 72$.

We can now use SPSS to determine the test statistic for this sample. Once the data file is open (see Section 1.3.1 for details on opening data files):

(1) Click on **Statistics** from the menu bar.

(2) Click on **Compare Means** from the pull-down menu.

(3) Click on **One-Sample T Test** from the pull-down menu to open the One-Sample T Test Dialog box (see Figure 11.2).

Figure 11.2 One-Sample T Test Dialog Box

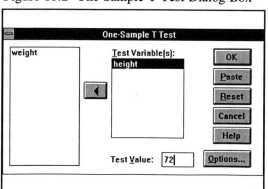

(4) Highlight the "height" variable and move it to the Test Variable(s) box by clicking on the **right arrow button.**

(5) In the Test Value box in the lower right-hand side of the dialog box, enter the number 72. (This value is represented as μ_0 in the textbook.)

(6) Click on **OK** to run the procedure.

The output should look like that in Figure 11.3.

Figure 11.3. Sample Output for One-Sample T-Test

```
One Sample t-tests

                          Number
Variable                  of Cases        Mean          SD       SE of Mean

HEIGHT                       56          73.7679       2.224        .297


       Test Value = 72

Mean                  95% CI
Difference      Lower      Upper      |   t-value       df       2-Tail Sig

    1.77         1.172      2.363     |    5.95         55          .000
```

From this listing we see that the mean height of the 56 football players is 73.77 inches and the standard deviation is 2.22 inches. The t-statistic for the test is t = (73.77 − 72) ÷ $2.22/\sqrt{56}$ = 5.95, which is also printed by the program.

The test statistic is compared to percentage points of the t-distribution with n − 1 = 55 degrees of freedom. From the table in the textbook, the significance points for a two-tailed test at the 5% level are approximately −2.01 and 2.01. The null hypothesis is rejected; we conclude that the mean height exceeds 6 feet. (If we were making a one-tailed test at the 5% level, the absolute value of the significance point would be between 1.671 and 1.684.) Alternatively, we may use the P value printed by SPSS. This is explained in Section 11.4.

Relation of Two-Tailed Tests to Confidence Intervals

The relationship between two-tailed tests and confidence intervals is straightforward. If μ_0 lies within the (1−α) confidence interval for the mean of a variable, the null hypothesis would not be rejected at the α level of significance. Conversely, if μ_0 is not within the interval, then the null hypothesis is rejected.

Return to the output from the analysis of football players' heights. The 95% confidence interval for the *difference* between the sample mean and μ_0 (72 inches) is displayed in the lower left-hand corner. (Note: 95% is the default; you may change this value by clicking on the **Options button** of the One-Sample T Test dialog box.) If 0 is contained within the interval, then H_0 is accepted. In this example, the interval ranges from a difference of 1.17 inches to 2.36 inches; because 0 is not in this range, H_0 is rejected at the 5% level.

11.4 P Values: Another Way to Report Tests of Significance

The P value is the probability of obtaining a value more extreme than a given test statistic given that the null hypothesis is true. When a computer program such as SPSS prints an appropriate P value, it is no longer necessary to look up the significance point(s) in the normal or t-table. In this manual, we will interpret the P values whenever they are produced correctly by SPSS. When SPSS does not produce a P value for a given procedure, we show how to obtain significance point(s) and P value(s) from the corresponding probability distribution.

Test for a Mean when the Population Standard Deviation is Known

Because SPSS does not compute an appropriate test of significance when the population standard deviation is known, the reported P value is also inappropriate. Correct P values may be found by referring to the table in Appendix I of the textbook.

Test for a Mean when the Population Standard Deviation is Not Known

SPSS does perform the t-test for a mean when the population standard deviation is unknown. It also reports a two-tailed P value associated with this test. Figure 11.3 displays the output for the test of heights of football players. The column labeled "2-Tail Sig" is the P value for this test. You will notice that the P value is listed as .000. It is not possible for the P value to be precisely equal to 0. Rather, this indicates that the value is less than .0005 and appears as 0 because it is rounded to three significant digits. (If the actual value were greater than .0005 and less than .0015, it would be rounded up, and appear as .001 in the printout.)

This P value indicates that, if the null hypothesis were true (football players were, on average, 6 feet tall), then the probability of obtaining a test statistic with an absolute value of 5.95 or greater is less than .0005. This means that we would reject the null hypothesis for any α level greater than .0005. So, when we use SPSS to compute the test statistic and perform the hypothesis test, we would compare the P value to our predetermined α level. If $P < \alpha$, then the null hypothesis is rejected.

SPSS reports only P values for two-tailed tests. If we are performing a one-tailed test, we are concerned only with the upper (or lower) tail of the t-distribution. In this case, in order to obtain the correct achieved significance level, the P value must be divided by 2. To reject the null hypothesis when conducting a one-tailed test, this achieved significance level (P/2) must be less than α *and* the sample mean must be in the direction specified by the alternative hypothesis (H_1).

Refer again to the height example (Figure 11.3). Suppose we want to test the null hypothesis H_0: $\mu \leq 72$ against the alternative H_1: $\mu > 72$, using $\alpha = .05$. The P value reported by SPSS is $P < .0005$. We divide P by 2 and find that the achieved significance level is less than .00025. This is less than α (.05) *and* the sample mean (73.77) is greater than 72 inches. Thus, we reject H_0 and conclude that football players are, on average, taller than 6 feet.

11.5 Testing Hypotheses About a Proportion

We can test hypotheses about a proportion using the Binomial procedure in SPSS. Using the "fire.sav" data file, let us test the hypothesis that the proportion of white applicants is less than the national population percentage, which is approximately 74%. The hypotheses are H_0: $p \geq .74$ and H_1: $p < .74$. After you open the data file:

(1) Click on **Statistics** from the menu bar.

(2) Click on **Nonparametric Tests** from the pull-down menu.

(3) Click on **Binomial** from the pull-down menu to open the Binomial Test dialog box (Figure 11.4).

(4) Click on and move the "race" variable to the Test Variable List box using the **right arrow button**.

(5) Type .74 in the Test Proportion: box. This represents p_0.

(6) Click on **OK** to run the procedure.

Figure 11.4 Binomial Test Dialog Box

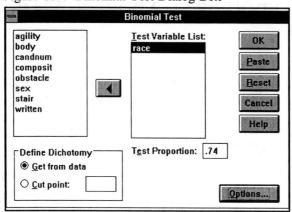

The output listing is displayed in Figure 11.5. The Test Proportion is the p_0 that you entered in step 5. The Obs. Prop. represents the proportion of cases with value of 1 in the data file (17/28). (Note that SPSS always calculates the proportion using the value of the variable with the larger number of cases. Therefore, you must be careful to enter the appropriate test proportion.)

The printout also contains the one-tailed P value[1] using the normal approximation to the binomial distribution. The continuity correction is employed to compute the z-statistic. Here, P = .0827. If we were using an α level of .05, we would accept the hypothesis that the proportion of white applicants is greater than or equal to .74. (Remember to check whether the observed proportion is above or below p_0 before deciding to accept or reject H_0.)

Figure 11.5 Sample Output for Binomial Test

```
- - - - - Binomial Test

  RACE

  Cases
                          Test Prop. =    .7400
      17      = 1         Obs. Prop. =    .6071
      11      = 2
      --                  Z Approximation
      28      Total       1-Tailed P =    .0827
```

---

[1]This is an exception to the SPSS rule that P values are printed for two-tailed tests.

11.6 Testing Hypotheses About a Median: The Sign Test

A hypothesis about the median of a variable may be conducted using the sign test. This is especially useful for ordinal variables or variables that have highly skewed distributions.

SPSS does not have an option on the menu bar that readily performs this sign test, but you can use it to find the number of observations below a specified value (M_0). This is efficient if the data set is very large.

We illustrate using the "noise.sav" data file to test the hypothesis that the median speed of cars on sections of a roadway is equal to 35 mph; the hypotheses are H_0: M = 35 mph and H_1: M ≠ 35 mph.

To compute the number of automobiles in the sample that traveled less than 35 mph, we need to tag these cases using the Count procedure, and then determine the number of tagged cases. To do this:

(1) Click on **Transform** from the menu bar.

(2) Click on **Count Occurrences** from the pull-down menu. This opens the Count Occurrences of Values within Cases dialog box (see Figure 11.6).

Figure 11.6 Count Occurrences of Values within Cases Dialog Box

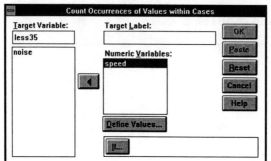

(3) In the Target Variable box, type the name of a new variable, e.g., "less35." This variable will contain a 1 for all the cases you wish to tag, and a 0 for all other cases. If you so desire, you may also include a variable label in the Target Label box.

(4) Click on the "speed" variable and move it to the Numeric Variables box using the **right arrow button**. This denotes the variable for which you want to count cases.

(5) Click on the **Define Values button** to open the Count Values within Cases: Values to Count dialog box (Figure 11.7). It is in this dialog box that you specify which values of the speed variable will be tagged.

Figure 11.7 Count Values within Cases: Values to Count Dialog Box

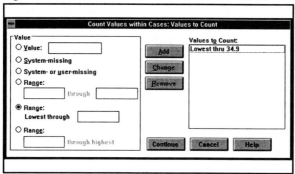

(6) We want to count all cases with values less than 35, so click on **Range: Lowest through_____** in the Value box.

(7) Type the value 34.9 in the space. (If you type 35, SPSS will count values of exactly 35).

(8) Click on the **Add button**. The line "lowest thru 34.9" should appear in the Values to Count box.

(9) Click on **Continue** to close the dialog box.

(10) Click on **OK** to run the procedure.

There will be a new variable in your data window that contains 1's and 0's. You must now count the total number of 1's in order to determine the total number of cases with values less than M_0. You can do so by requesting the sum of the "less35" variable from the Frequencies procedure. You should obtain a sum of 11.

You can now compute the test statistic $z = [(11/30) - .5]/[.5 \times (1/\sqrt{30})] = -1.461$. Significance points for a two-tailed test at the 5% level, from the standard normal distribution, are -1.96 and 1.96. We conclude that the data are consistent with a median speed of 35 mph.

11.7 Paired Measurements

11.7.1 Testing Hypotheses About the Mean of a Population of Differences

SPSS will conduct a test of whether the mean of a population of difference scores is equal to 0. We illustrate with the "reading.sav" data file, based on the data in Table 10.2 of the textbook. This file contains reading scores for 30 students obtained on the same test administered before and after second grade. We want to determine whether reading skill increases, on average, throughout second grade. After opening the data file:

(1) Click on **Statistics** from the menu bar.

(2) Click on **Compare Means** from the pull-down menu.

(3) Click on **Paired-Samples T Test** from the pull-down menu. This opens the Paired-Samples T Test dialog box (see Figure 11.8).

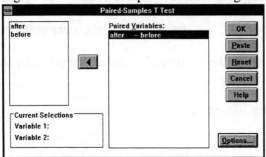

Figure 11.8 Paired-Samples T Test Dialog Box

(4) Click on the "after" variable. It will appear in the Current Selections box as Variable 1.

(5) Click on the "before" variable. It will appear in the Current Selections box as Variable 2.

(6) Move the names of the variables into the Paired Variables box by clicking on the **right arrow button**.

(7) Click on **OK** to run the procedure.

Figure 11.9 displays the output from this procedure. The top portion lists the means, standard deviations, and standard errors of reading scores before and after second grade. The scores increased, on average, from 1.52 to 2.03. This portion also includes the sample correlation coefficient between the pre-test and post-test (.573) and a test of significance of the correlation (P = .001).

The lower portion contains information regarding the test of the hypothesis that the mean difference is equal to 0. The mean difference, .5100 is equivalent to 2.0333 − 1.5233. The table also displays the standard deviation and the standard error of the difference. The t-statistic is $t = (.5100 - 0)/.492/\sqrt{30} = 5.68$, which is also printed. A decision about H_0 can be made by reference to the t-distribution with 29 degrees of freedom or by using the P value displayed by SPSS. The 2-tail Sig is the P value (< .0005).

This P value pertains to a two-tailed test, however. If we began by speculating that reading scores would increase, we would conduct a one-tailed test, with the alternative hypothesis $H_1: \mu_d > 0$. Therefore, we must compare the P value to 2α and verify that the sample post-test mean is higher than the sample pre-test mean. Doing so, we would reject H_0 at most reasonable α levels.

Figure 11.9 Sample Output for T-Test for Paired Samples

```
t-tests for Paired Samples


                Number of           2-tail
Variable          pairs      Corr    Sig         Mean          SD         SE of Mean

AFTER     Reading Score After 2nd Grade         2.0333        .594           .109
                    30         .573    .001
BEFORE    Reading Score Before 2nd Grade        1.5233        .276           .050



            Paired Differences           |
    Mean         SD       SE of Mean     |    t-value         df         2-tail Sig

    .5100       .492         .090        |     5.68           29           .000
    95% CI (.326, .694)                  |
```

Note: Due to a small error in calculation, the standard deviation in the textbook for this problem is incorrect.

11.7.2 Testing the Hypothesis of Equality of Proportions

You may also examine changes in time for dichotomous variables by testing the equality of two proportions obtained from a single sample. The textbook refers to the test of whether the two proportions are equal as a test for a "turnover table," and SPSS labels it the McNemar test for correlated proportions. Using SPSS, the procedure produces a chi-square test statistic, but the conclusions are equivalent to those from the z-test discussed in the textbook.

We will illustrate this using the "war.sav" file. This file contains the data from the study discussed in Exercise 11.25 of the textbook. This study examined changes in attitudes regarding the likelihood of war. In both June and October of 1948, subjects were asked to indicate whether or not they expected a war in the next ten years. The "war.sav" data file is coded so that a 2 represents "Expects War" and a 1 represents "Does Not Expect War."

To test the hypothesis of equal proportions, open the data file and:

(1) Click on **Statistics** from the menu bar.

(2) Click on **Nonparametric Tests** from the pull-down menu.

(3) Click on **2 Related Samples** from the pull-down menu. This opens the Two-Related-Samples Tests dialog box (Figure 11.10).

Figure 11.10 Two-Related-Samples Test Dialog Box

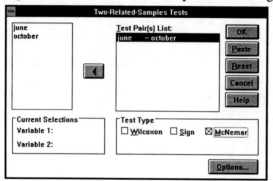

(4) Click on the variable name "june." It will appear as Variable 1 in the Current Selections box.

(5) Click on the variable name "october." It will appear as Variable 2 in the Current Selections box.

(6) Move the paired variables to the Test Pair(s) List box by clicking on the **right arrow button**.

(7) In the Test Type box, click off the **Wilcoxon** box and click on the **McNemar** box.

(8) Click on **OK**.

Figure 11.11 McNemar Test for Likelihood of War

```
- - - - -  McNemar Test

     JUNE         june
with OCTOBER     october


                         OCTOBER
                    1.00        2.00           Cases             597
                    ---------------------
             1.00 |   194  |     45  |         Chi-Square     53.1302
      JUNE        ---------------------
             2.00 |   147  |    211  |         Significance =   .0000
                    ---------------------
```

The output (Figure 11.11) displays the data in the two-way frequency table identical to that in Table 11.19 of the textbook. From it we see that 45 people who did not think there would be war when questioned in June changed their minds for the October polling. The listing also reports the chi-square test statistic (53.1302) and the corresponding P value (P < .00005). Therefore, we would reject the null hypothesis of equal proportions at any reasonable α level.

Note that the McNemar test is a two-tailed test. To perform a one-tailed test, compare P/2 to α and check that the sample proportions are in the direction indicated by the alternative hypothesis.

Chapter Exercises

11.1 Use SPSS and the "noise.sav" data file to complete the following:

(a) Test the null hypothesis that the average speed of automobiles is equal to 35 mph. Use $\alpha = .05$, and state the test statistic and conclusion.
(b) Compare the result in (a) to that obtained using the sign test (Section 11.6).
(c) Which test would you perform? On what did you base your decision?

11.2 Use SPSS and the "head.sav" data file, containing information on length and breadth of head measurements for a sample of infants, to complete the following:

(a) Find the 90% confidence interval for the length of infants' heads. State the sample mean.

(b) Would you reject the null hypothesis H_0: $\mu = 181$ cm, using $\alpha = .10$? (Hint: refer to your conclusions in part (a).) Using $\alpha = .01$?
(c) What are your conclusions for the hypothesis H_0: $\mu \leq 181$ cm, using $\alpha = .10$?
(d) What is the P value for the hypothesis in part (b)? in part (c)?

11.3 Use SPSS and the "gas.sav" data file, which contains tax rate (in cents per gallon) for the 50 states, to complete the following:

(a) What is the P value for the test of the null hypothesis that the gas tax rate is, on average, equal to 18.5 cents per gallon?
(b) Would you reject the hypothesis at $\alpha = .05$? $\alpha = .01$?

11.4 Use the "semester.sav" data file to complete the following:

(a) Estimate the proportion of students who are majoring in engineering, statistics, or materials science compared to those majoring in psychology, public administration, architecture, or industrial administration. (Hint: You need to recode the "major" variable.)
(b) Test the hypothesis that the proportion is greater than 50%. Use $\alpha = .05$.
(c) Using $\alpha = .10$, test the hypothesis that the proportion majoring in engineering/statistics/materials science is less than 50%.
(d) Find the 95% confidence interval for the mean of semester grades.
(e) Referring to semester grades, test the hypothesis H_0: $\mu = 230$ at the .05 level of significance. State the P value.
(f) Repeat part (e), this time testing the one-tailed hypothesis H_0: $\mu \leq 230$.

11.5 Use the "weather.sav" data file to complete the following. This data file contains weather information for large U.S. cities (cities with a population over 200,000 in 1992). There are two variables: average temperature on July 2nd, and average annual rainfall (in inches). These averages were determined based data from the 30 year period from 1961 to 1990.

(a) What is the mean rainfall for large cities? The median rainfall?
(b) Would you reject the hypothesis that the median is equal to 40, using $\alpha = .025$?
(c) Test the hypothesis that the mean rainfall in cities is not equal to 40.
(d) Explain the difference, if any, in your results for parts (b) and (c).

11.6 Using the "conform.sav" data file, answer the following questions:

(a) Are wives more comformist, on average, than their husbands (use $\alpha = .05$)?
(b) What is the minimum α for which you would reject the null hypothesis indicated in part (a)?

Appendix XI
SPSS Syntax for Answering Questions About Population Characteristics

11.1.2 Validity Conditions

    Use the following syntax to create a histogram of the "lang" variable imposing the normal curve on the graph:

    Frequencies variables=lang
      /format=notable
      /histogram=normal.

11.5 Testing Hypotheses About a Proportion

    Use the following command to test whether the proportion of white applicants is less than .74. The race(1) portion of the command defines the cutoff value for the race variable. So, all values less than and equal to 1 are combined into one category, and all those greater than 1 are placed in the other category.

    Npar tests binomial (.74) = race (1).

11.6 Testing Hypotheses About a Median: The Sign Test

    Use the first command below to compute a variable named "less35" that has a value of 1 if the value of the "speed" variable is less than 35, and 0 otherwise. Use the second command to compute the total of the "less35" variable.

    Count less35=speed (lowest thru 34.9).
    Frequencies variables=less35
        /statistics=sum.

11.7.1 Testing Hypotheses About the Mean of a Population of Differences

    Use the syntax below to test the hypothesis about the means of a population of differences for variables "before" and "after:"

    T-test pairs=after before.

Chapter 12 DIFFERENCES BETWEEN TWO POPULATIONS

This chapter focuses on the comparison of two independent populations. SPSS does not have a procedure for testing hypotheses about means when the population standard deviations are known, or any procedure for testing hypotheses about two proportions; these tests must be computed by hand. There are procedures in SPSS for comparing means when the standard deviations are estimated from the sample, however.

12.1 Comparison of Two Independent Sample Means when the Population Standard Deviations are Known

Although SPSS does not conduct a test comparing means of two independent samples when the standard deviations (σ's) are known, we can use SPSS to find the sample means and calculate the test statistic by hand.

Suppose, for example, that we had reason to believe that students who take the SAT in the fall do better, on average, on the verbal portion of the test than do those who take it in the spring. The null and alternative hypotheses for this test are H_0: $\mu_1 \leq \mu_2$ and H_1: $\mu_1 > \mu_2$, where population 1 is fall test-takers and population 2 is spring test-takers. (Assume that no student takes the test on both occasions.)

Suppose that we had a data set containing scores of a random sample of 1,000 students who took the exam in the fall, and of 750 students who took it in the spring. Using the Frequencies, Descriptives, or Explore procedure, we could find the mean for each group. Suppose further that the average score for the fall sample is 520, for the spring sample, it is 482.

We know that the standard deviation for the SAT is preset by the test publisher to 100. The test statistic is z = (520 − 482) / $\sqrt{(100^2/1000 + 100^2/750)}$ = 7.867. The significance point for the one-tailed test using α = .01 is approximately 2.33; the null hypothesis is rejected.

SPSS also does not readily calculate confidence intervals when the population standard deviations are known. Again, you must compute these by hand using formula 12.2 in the textbook.

12.2 Comparison of Two Independent Sample Means when the Population Standard Deviations are Unknown but Treated as Equal

On many occasions, the standard deviations of the populations are unknown and must therefore be estimated. Testing procedures are slightly different based on whether or not we can assume that the two standard deviations are equal. SPSS performs the test for both types of conditions. This section discusses the more common situation in which the variances are treated as equal.

The "final.sav" data file, based on Table 12.4 of the textbook, contains information on final grades in statistics for 68 students: 39 undergraduates and 29 graduates. Suppose we wish to determine whether graduates and undergraduates differ, on average, in their final course grade. The null and alternative hypotheses are $H_0: \mu_1 = \mu_2$ and $H_1: \mu_1 \neq \mu_2$, where population 1 = undergraduate, and population 2 = graduate. We shall test the hypothesis at the 5% level.

We can now use SPSS to conduct the test as follows. After opening the data file:

(1) Click on **Statistics** from the menu bar.

(2) Click on **Compare Means** from the pull-down menu.

(3) Click on **Independent Samples T-Test** from the pull-down menu to open the Independent-Samples T Test dialog box (Figure 12.1).

Figure 12.1 Independent-Samples T Test Dialog Box

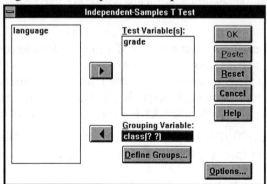

(4) Click on and move the "grade" variable the Test Variable(s) box using the **upper right arrow button**.

(5) Click on and move the "class" variable to the Grouping Variable box using the **lower right arrow button**.

(6) Notice that two question marks appear in parentheses after the variable "class." This signifies that you need to indicate the two values of the class variable for which you wish to calculate mean differences. To do so, click on **Define Groups** to open the Define Groups dialog box.

Figure 12.2 Define Groups Dialog Box

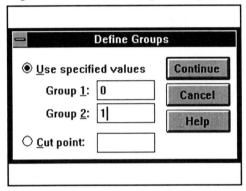

(7) Figure 12.2 shows the Define Groups dialog box. In our example, undergraduates are coded 0, and graduates are coded 1. Therefore, enter these numbers in the Group 1 and Group 2 box. (The cut point option is used if there more than two values of the grouping variable.)

(8) Click on **Continue** to close the dialog box.

(9) Click on **OK** to run the procedure.

The output is displayed in Figure 12.3. The upper portion of the listing displays summary information (n's, means, standard deviations, and standard errors) for each of the samples. In this course, undergraduates had an average final grade of 136.44 points, and graduates had an average of 127.00 points. The difference is 136.44 − 127.00 = 9.44 points.

The lower portion displays two different t-statistics, one based on the assumption of equal variances, the other assuming unequal variances. We will only consider the equal variances case. The test statistic t = 9.44/5.13 = 1.84, where 5.13 is the standard error of the difference assuming equal population variances.

The t-statistic is compared to significance points from the t-distribution with 68 − 2 = 66 degrees of freedom. This is done by SPSS resulting in the printed P value. Since P = .071 is greater than .05, the null hypothesis is accepted.

Figure 12.3 T-Test for Independent Samples

```
t-tests for Independent Samples of CLASS     class
```

Variable	Number of Cases	Mean	SD	SE of Mean
GRADE				
undergra	39	136.4359	20.230	3.239
graduate	29	127.0000	21.866	4.061

Mean Difference = 9.4359

Levene's Test for Equality of Variances: F= .068   P= .795

	t-test for Equality of Means				95%
Variances	t-value	df	2-Tail Sig	SE of Diff	CI for Diff
Equal	1.84	66	.071	5.134	(-.815, 19.687)
Unequal	1.82	57.75	.074	5.194	(-.963, 19.834)

The output also includes a 95% confidence interval for the mean difference. That is, the difference between graduates' and undergraduates' average scores are between −0.82 and 19.69 points with 95% confidence. Because 0 is in this range, the results of the significance test are confirmed.

One-Tailed Tests

The SPSS procedure for conducting a one-tailed test is the same as that for a two-tailed test but differs in how the P value is used. Because the P value reported is for a two-tailed test, we must compare the P to 2α, and also verify that the sample means differ in the direction supported by the alternative hypothesis. In the example, .071 < .10, and the sample mean for undergraduates is greater than the sample mean for graduates. Thus, we would reject H_0: $\mu_1 \leq \mu_2$ in favor of H_1: $\mu_1 > \mu_2$ at the 5% level of significance.

12.3 Comparison of Two Independent Sample Means when the Population Standard Deviations are Unknown and not Treated as Equal

Figure 12.3 also shows an approximate t-statistic, estimated number of degrees of freedom, P value, and 95% confidence interval for the case when the population standard deviations are not considered to be equal. These do not correspond to the procedures described in the textbook. The reader is referred to Section 12.3 for alternative approaches that are recommended when variances are not equal.

Other Information in the Output

The output in Figure 12.3 also contains the results for Levene's test for the equality of two variances. The null hypothesis is that the variances of the two populations are equal. This procedure is an alternative to the F-test for equal variances described in the textbook. The reader is referred the Section 13.2 for the recommended approach.

12.4 Comparison of Two Independent Sample Proportions

We shall describe how to use SPSS to test the equality of two proportions using data from the "meal.sav" data file. In this application we test whether the proportion of pasta (vs. French/seafood) restaurants is greater among chain restaurants than among non-chain restaurants.

SPSS does not have a procedure for conducting the test using the z-statistic and standard normal distribution[1]. However, we can use SPSS to create the two-way frequency table, and then compute the test statistic by hand.

After opening the "meal.sav" data file, we must recode the meal variable, combining French (value=2) and seafood (value=3) into one group. The recoding procedure is described in Section 1.4.2, but an outline of the steps follows:

(1) Click on **Transform** from the menu bar.

---

[1] SPSS will compute a chi-square test of independence. As discussed in Section 14.2 of the textbook, both tests are equivalent.

(2) Click on **Recode** from the pull-down menu.

(3) Click on **Into Different Variable** from the pull-down menu.

(4) Highlight the "meal" variable and move it to the Input Variable --> Output Variable box by clicking on the **right arrow button.**

(5) In the Output Variable box, type the name of the new variable (e.g., "pasta"), and click on **Change**.

(6) Click on the **Old and New Values button**.

(7) In the Old Values box, enter 1 in the Value option; in the New Values box, click on the **Copy Old Value(s)** option.

(8) Click on **Add**.

(9) In the Old Values box, click on the **All Other Values** option, and in the New Values box, enter 2 in the Value option.

(10) Click on **Add**.

(11) Click on **Continue**.

(12) Click on **OK**.

There should be a new variable called "pasta" in your data window, with a value of 1 for pasta restaurants, and 2 for other.

Now, we need to create the two-way frequency table. The procedure for creating crosstabulation tables is detailed in Chapter 6, but a brief outline follows:

(1) Click on **Statistics** from the menu bar.

(2) Click on **Summarize** from the pull-down menu.

(3) Click on **Crosstabs** from the pull-down menu.

(4) Move the "pasta" variable to the row box, and the "chain" variable to the column box by clicking on the arrow buttons.

(5) Click on the **Cells button**.

(6) In the Percentages box, click on the **Column** option.

(7) Click on **Continue**.

(8) Click on **OK**.

The resulting crosstabulation table is displayed in Figure 12.4.

Figure 12.4 Crosstabulation Table

```
PASTA   by  CHAIN   chain restaurant

                    CHAIN           Page 1 of 1
            Count   |
            Col Pct |yes       no
                    |                    Row
                    |     1.00|     2.00| Total
PASTA       --------------------------------
            1.00    |     8   |     7   |    15
                    |    36.4 |    29.2 |   32.6
                    --------------------
            2.00    |    14   |    17   |    31
                    |    63.6 |    70.8 |   67.4
                    --------------------
            Column        22        24        46
            Total       47.8      52.2     100.0

Number of Missing Observations:  0
```

We see that the proportions of chain and non-chain restaurants that are pasta restaurants are 0.364 and 0.292, respectively. The proportion of all restaurants that specialize in pasta is 0.326. The test statistic is $z = (0.364 - 0.292)/\sqrt{0.326 \times 0.674 \times (1/22 + 1/24)} = 0.520$, using the equation in Section 12.4 of the textbook. This test statistic is very small, so H_0 is accepted at any reasonable α level; we conclude that the proportion of pasta restaurants is the same regardless of chain status.

12.5 The Sign Test for a Difference in Locations

The sign test is used to test differences between the medians of two independent samples. SPSS does have a procedure called the Sign Test, but it is somewhat different from that described in the textbook. Therefore, we will use SPSS to create the crosstabulation table, and compute the test statistic by hand.

Since all values exactly equal to the median are discarded for purposes of determining the test statistic, we will first have to identify the common median, and then define it as a missing value. We will illustrate using data from the "salary.sav" data file. This file contains salaries for a sample of men and women; we will examine the gender differences in earnings. The median salary for all individuals in the sample can be obtained with the Descriptives procedure. The common median is 94, or $9,400.

In order to define this value as missing:

(1) In the Data window, click on (or move your cursor to) the salary variable.

(2) Click on **Data** from the main menu.

(3) Click on **Define Variable** from the pull-down menu. This will open the Define Variable: salary dialog box.

(4) Click on the **Missing Values button** in the Change Settings box. This opens the Define Missing Values: salary dialog box.

(5) Click on the **Discrete Missing Values**.

(6) Type 94 in the first box.

(7) Click on **Continue** to close the dialog box.

(8) Click on **OK** to run the procedure.

The value 94 is now a user-defined missing value. Before obtaining the two-way frequency table, we must create a dichotomous variable with the value 1 for all individuals with a salary below the median, and the value 0 for those with salaries above the median. The recoding procedure is described in Section 1.4.2, but the following is a brief outline:

(1) Open the Recode into Different Variable dialog box.

(2) Indicate that the "salary" variable is to be recoded into the new variable called "medsplit."

(3) Set up the recoding criteria to recode (a) lowest thru 93 into 1, (b) 95 thru highest into 0, and (c) user missing into system missing.

(4) Run the recode procedure by clicking on **OK**.

All that remains is to obtain the crosstabulation of "medsplit" and "gender." The column variable should be gender, and the row variable medsplit. You should also request column percentages (see Section 6.1.1). Your results should resemble those in Figure 12.5.

Figure 12.5 Crosstabulation Table

```
MEDSPLIT   by   GENDER   gender

                    GENDER           Page 1 of 1
              Count  |
              Col Pct |male        female
                     |                          Row
                     |    1.00|     2.00|     Total
MEDSPLIT      -------------------------------
          .00  |      11    |      5    |    16
                |    61.1    |    35.7    |  50.0
                -------------------------------
         1.00  |       7    |      9    |    16
                |    38.9    |    64.3    |  50.0
                -------------------------------
         Column         18         14          32
         Total        56.3       43.8       100.0

Number of Missing Observations:   1
```

Notice the advisory note in this output indicating that there is one missing observation. The table itself is identical to Table 12.9 in the textbook. You must complete the remaining steps of the Sign Test by hand, in the manner discussed in Section 12.5 of the textbook.

Chapter Exercises

12.1 Use the "enroll.sav" data file and SPSS to test whether there is a significant difference in the racial disproportion index between districts with high and low percentages of students paying full price for lunch, as follows:

(a) Recode the "pct_lnch" variable into a dichotomous variable, with 52% as the split point. (That is, values less than 52% will constitute the "low" group, and all other values the "high" group.)
(b) Would you perform a one- or two-tailed test? Why?
(c) Based on your response to part (b), state the null and alternative hypothesis and the significance point(s) (using $\alpha = .05$).
(d) Use SPSS to conduct the test. State the value of the test statistic and the P value. (Assume equal variances.) Is H_0 rejected if $\alpha = .05$? If $\alpha = .01$?

12.2 Use the "final.sav" data file to complete the following:

(a) State the null and alternative hypotheses for testing whether, on average, native English speakers perform better in statistics than do non-native English speakers.
(b) Assuming equal variances, use SPSS to conduct the test. What are your conclusions using $\alpha = .01$? Give the P value for the test.
(c) Interpret the 99% confidence interval for the mean difference between populations. (Hint: 99% is not the default option for confidence intervals in SPSS.)
(d) Suppose that the assumption of equal variances is not reasonable. Would your conclusions differ?

12.3 Use SPSS and the "final.sav" data to test whether the proportion of native English speakers who received a final statistics grade of over 130 points is equal to the proportion of non-native English speakers receiving such a grade.

(a) State the null and alternative hypotheses.
(b) What are the values of the sample proportions? Of the test statistic?
(c) What are your conclusions?

12.4 Your local common council is interested in reducing noise pollution that results from speeding automobiles in your neighborhood. Using the "noise.sav" data file, test their assumption that this can be accomplished by strictly enforcing a 40 mph speed limit, as follows:

(a) Recode the speed variable, using 40mph as the cut point, and conduct the sign test to compare the median noise level of cars going under 40 mph with those going over 40 mph. Use $\alpha = .05$.
(b) Prepare a 1-2 sentence "report" to the common council outlining your findings and recommendations.

Appendix XII
SPSS Syntax for Differences Between Two Populations

12.2/12.3 Comparison of Two Independent Sample Means when the Population Standard Deviations are Unknown

Use the following command to compute a t-test for the "grade" variable based on "class:"

T-test groups=class(0,1)
    /variables=grade.

12.5 The Sign Test for a Difference in Locations

Use the following syntax to perform the sign test for "salary" based on gender:

Npar tests median = salary by gender(1,2).

Chapter 13 VARIABILITY IN ONE POPULATION AND IN TWO POPULATIONS

This chapter focuses on tests about population variances. Two major tests are discussed, those concerning one population, and those involving two independent populations. SPSS does not have a procedure for the one-sample case. It does provide a test for the two sample case, but this test is different than the one discussed in the textbook. Therefore, for both cases, we will use SPSS to determine the sample variances, and then calculate the test statistics by hand.

13.1 Variability in One Population

13.1.1 Testing the Hypothesis that the Variance Equals a Given Number

We illustrate a statistical test for one variance using the weights of 25 dieters in the "dieter.sav" data file. Suppose we know that the standard deviation of the weight of American adults is 9 pounds and wish to examine whether the variance for overweight individuals is also equal to 81.

The procedure to test this hypothesis is as follows:

(a) State the null and alternative hypotheses, choose a significance level, and find the significance point on the chi-square distribution. In the example, the hypotheses are H_0: $\sigma^2 = 81$ and H_1: $\sigma^2 \neq 81$; we shall use the 5% significance level.

The significance points are taken from the χ^2 distribution with $n - 1 = 24$ degrees of freedom. From Appendix IV in the textbook, we see that $\chi^2_{24}(.025) = 39.364$, and $\chi^2_{24}(.975) = 12.401$; we will reject H_0 if the value of the test statistic is less than 12.401 or greater than 39.364.

(b) Use SPSS to find the variance of the sample, and then compute the test statistic. In this example, the variance, determined from the Explore procedure in SPSS, is 76.707. The test statistic is $\chi^2 = [(n-1)s^2]/\sigma_0^2 = (24 \times 76.707)/81 = 22.728$. Because 22.728 is not in the region of rejection, we conclude that the variance of weights of overweight adults does not differ from 81. The P value interval for the test is $P > .10$.

13.1.2 Confidence Intervals

SPSS does not have a procedure to calculate the confidence interval for a variance. The expression for an interval with confidence coefficient $1-\alpha$ is:

$$[(n-1)s^2/\chi^2_{(n-1)}(\alpha/2), \quad (n-1)s^2/\chi^2_{(n-1)}(1-\alpha/2)]$$

For the example with overweight adults, the 95% confidence interval is:

$$[(24 \times 76.707)/39.364, \quad (24 \times 76.707)/12.401] = (46.77, \quad 148.45)$$

As expected, the confidence interval contains the value 81.

13.2 Variability in Two Populations

13.2.1 Testing the Hypothesis of Equality of Two Variances

We illustrate the statistical test for two variances using the obstacle course times of 14 male and 14 female firefighters in the data file "fire.sav." Suppose, for example, we suspect that the obstacle course scores of male firefighter applicants are more homogenous than those of female applicant, that is, we raise the question of whether males perform rather consistently on the obstacle course, but that there is a wider range of performance times among women.

The procedure to test this hypothesis is as follows:

(a) State the null and alternative hypotheses, choose a significance level, and find the significance point on the F-distribution. In the example, the hypotheses are $H_0: \sigma_1^2 \geq \sigma_2^2$ and $H_1: \sigma_1^2 < \sigma_2^2$, where population 1 = male, and population 2 = female.

We shall use the 5% significance level. The significance point is taken from the F-distribution with $n_1 - 1 = 13$ degrees of freedom in the numerator and $n_2 - 1 = 13$ degrees of freedom in the denominator. The significance point is $F_{13,13}(.975) = 1/F_{13,13}(.025) = 1/3.1532$. Therefore, we will reject the null hypothesis if the test statistics is less than .317.

(b) Use SPSS to obtain the sample variances and compute the test statistic by hand. There are several methods that may be used to obtain standard deviations for subgroups of data in SPSS. You could, for instance, use the Select if command (see Section 1.4.3) and Descriptives procedure (Section 3.3) to obtain descriptive statistics separately for males and females. Or, you could use the Explore procedure as follows:

(1) Click on **Statistics** from the menu bar.

(2) Click on **Summarize** from the pull-down menu.

(3) Click on **Explore** from the pull-down menu.

(4) Click on and move the "obstacle" variable to the Dependent List box using the **upper right arrow button**.

(5) Click on and move the "sex" variable to the Factor List box using the **middle right arrow button**.

(6) Click on the **Statistics** from the Display box.

(7) Click on **OK** to run the procedure.

The listing should look like that in Figure 13.1. It contains two tables of descriptive statistics, one for males and one for females. These tables indicate that the variance of course times for males is $s_1^2 = 160.14$, and for females it is $s_2^2 = 570.56$. The test statistic is $F = s_1^2/s_2^2 = 160.14/570.56 = 0.281$. The test statistic is less than 0.317 and falls in the region of rejection; we conclude that males are more homogenous than females with regard to performance on the obstacle course.

Figure 13.1 Descriptive Statistics for Obstacle Course Times by Sex

```
      OBSTACLE
By    SEX        1           male

Valid cases:        14.0   Missing cases:        .0   Percent missing:        .0


Mean        96.1786   Std Err        3.3821   Min        82.3000   Skewness    1.2839
Median      92.7000   Variance     160.1387   Max       128.0000   S E Skew     .5974
5% Trim     95.1817   Std Dev       12.6546   Range      45.7000   Kurtosis    1.8501
95% CI for Mean (88.8720, 103.4851)           IQR        17.4500   S E Kurt    1.1541


      OBSTACLE
By    SEX        2           female

Valid cases:        14.0   Missing cases:        .0   Percent missing:        .0


Mean       132.8786   Std Err        6.3839   Min       100.7000   Skewness     .5681
Median     127.6000   Variance     570.5649   Max       178.8000   S E Skew     .5974
5% Trim    132.1151   Std Dev       23.8865   Range      78.1000   Kurtosis    -.6600
95% CI for Mean (119.0869, 146.6702)          IQR        37.1750   S E Kurt    1.1541
```

13.2.2 Confidence Intervals for the Ratio of Two Variances

SPSS does not have a procedure for confidence intervals for a variance ratio. Thus, you can use SPSS to determine the variances of the two independent samples and then compute the interval by hand as described in Section 13.2 of the textbook.

Chapter Exercises

13.1 Using the "final.sav" data file, determine the following:

(a) State the null and alternative hypothesis for testing whether there is a difference in variation of final grade scores between undergraduates and graduates.
(b) Select an α level and state the corresponding significance point(s).
(c) Use SPSS to obtain the sample variances. Compute the test statistic and state the conclusion of the hypothesis test.
(d) What are the upper and lower bounds of the confidence interval with coefficient $1-\alpha$?

13.2 Use the "salary.sav" data file to test the hypothesis that women's salaries are less variable than are men's. Use $\alpha=.01$. State the null and alternative hypotheses, give the significance point(s), your conclusion, and the P value for the test statistic.

13.3 Use SPSS and the "gas.sav" data file to test the following hypotheses. For both, state the null and alternative hypotheses, the α level, the significance point(s), and your conclusion.

(a) Is the average gas tax less than 20 cents?
(b) Is the standard deviation of gas tax rates equal to 5 cents?

Appendix XIII
SPSS Syntax for Variability in One Population and in Two Populations

13.2.1   Testing the Hypothesis of Equality of Two Variances

Use the following syntax to compute descriptive statistics for the variable "obstacle" separately for males and females:

means tables = obstacle by sex.

PART V

STATISTICAL METHODS FOR OTHER PROBLEMS

Chapter 14 INFERENCE ON CATEGORICAL DATA

This chapter describes how to use SPSS for test of goodness of fit with equal and unequal probabilities, to perform a chi-square test of independence, and to calculate measures of association such as the phi coefficient, coefficient lambda, and coefficient gamma.

14.1 Tests of Goodness of Fit

Equal Probabilities

In this illustration, we will use the data in Table 14.1 of the textbook test the hypothesis that there are equal probabililties of death occuring in one's birth month or any other month of the year. The null hypothesis is H_0: $p_1 = p_2 = \ldots = p_{12} = 1/12$. The data are contained in the file "death.sav" with n = 348 lines. Each entry is a number indicating the individual's month of death relative to the month of birth; for example, -6 indicates that the month of death is 6 months prior to the month of birth, 0 indicates that both months are the same, and so on. (For running the procedure with SPSS, any 12 different values could be used.)

By default, SPSS calculates the chi-square statistic to test the hypothesis of equal proportions. After you have opened the data file:

(1) Click on **Statistics** from the menu bar.

(2) Click on **Nonparametric tests** from the pull-down menu.

(3) Click on **Chi-Square** to open the Chi-Square Test dialog box (see Figure 14.1).

(4) Click on the variable name ("month") and the **right arrow button** to move it into the Test Variable List box.

(5) Click on **OK**.

The output should appear as shown in Figure 14.2.

Figure 14.1 Chi-Square Test Dialog Box

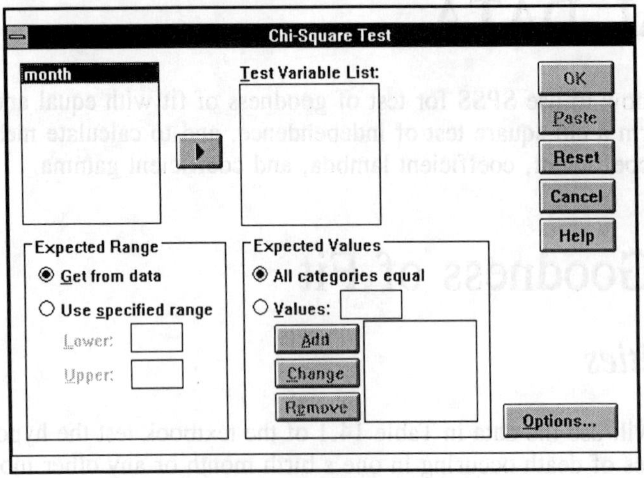

Figure 14.2 Chi-Square Test of Goodness of Fit

```
- - - - - Chi-Square Test
```

```
    MONTH
                Cases
    Category  Observed   Expected   Residual

         -6        24       29.00      -5.00
         -5        31       29.00       2.00
         -4        20       29.00      -9.00
         -3        23       29.00      -6.00
         -2        34       29.00       5.00
         -1        16       29.00     -13.00
          0        26       29.00      -3.00
          1        36       29.00       7.00
          2        37       29.00       8.00
          3        41       29.00      12.00
          4        26       29.00      -3.00
          5        34       29.00       5.00
                  ---
    Total        348

      Chi-Square              D.F.         Significance
        22.0690                11                .0238
```

Since we hypothesized that each month is equal in proportion, we see that the expected number of individuals who died during each of the 12 months is 348/12 = 29. The test statistic is 22.07 with 11 degrees of freedom. The P value of .0238 would lead us to reject H_0 at the 5% level (but not at the 1% level).

Probabilities Not Equal

It is also possible to conduct goodness-of-fit tests when the proportions are not hypothesized to be equal in the categories. For example, the current national percentage of white (nonminority) individuals in the United States is about 76%. We will use the data in "fire.sav" to test whether the proportion of white firefighter applicants is also equal to 76%, and thus is typical of the American population as a whole. The null hypothesis is H_0: $p_1 = .76$, $p_2 = .24$. To accomplish this: open the "fire.sav" data file and follow steps 1-4 above, clicking on the variable "race." Then:

(1) Click on **Values** in the Expected Values box.

(2) Enter the value (proportion) that you hypothesize for the first category of your variable. In this example, white is coded "1" and minority is coded "2." Since we are hypothesizing that the proportion of whites is equal to 76%, enter .76 in the Value box and click on **Add**.

(3) Enter the value for the next category of your variable. In this example, enter .24, since we hypothesize that the proportion of minorities is equal to 24%. Click on **Add**.

(4) After entering all of the expected values, click on **OK**.

The output should appear as shown in Figure 14.3.

Figure 14.3 Goodness of Fit Test for Unequal Proportions

```
- - - - - Chi-Square Test

    RACE
                                  Cases
                     Category  Observed   Expected   Residual

       white              1          17      21.28      -4.28
       minority           2          11       6.72       4.28
                                      --
                     Total          28


           Chi-Square              D.F.         Significance
              3.5868                 1                .0582
```

The Expected column gives you the hypothesized number of cases. For example, 76% of the sample of 28 individuals would be 21.28. The actual number of whites in the sample was 17 (see the Cases Observed column). The test statistic is $X^2 = 3.59$ and is not significant at the 5% level (P = .0582). Thus, we conclude that the sample of firefighter applicants is not different from the national white/minority distribution.

14.2 Chi-Square Tests of Independence

In Chapter 6 we demonstrate how to calculate frequency tables using the Crosstabs command. In this section, we will test the hypothesis that in a population two variables are independent on the basis of a sample drawn from that population. SPSS readily computes the appropriate X^2 test statistic.

To illustrate, we shall use the data in Table 12.17 of the text ("computer.sav") that shows the relationship between the occupational type and gender of individuals in computer magazine advertisements. Consider the hypothesis that occupational type is independent of gender in the population of advertisements represented by this sample. (Note that since there are just two gender groups, this is the same as a hypothesis that the probability of an occupational portrayal as seller, manager, clerical, computer expert, or other is the same for males and females.)

To test this hypothesis, open the "computer.sav" data file and follow these steps:

(1) Click on **Statistics** from the menu bar.

(2) Click on **Summarize** from the pull-down menu.

(3) Click on **Crosstabs** to open the Crosstabs dialog box.

(4) Click on the name of the row variable ("gender") and the **right arrow button**.

(5) Click on the name of the column variable ("role") and the **right arrow button**.

(6) Click on the **Cells** button to open the Crosstabs: Cell Display dialog box.

(7) Click on **Row** and **Column** in the Percentages box to indicate that you want row and column percentages.

(8) Click on **Continue**.

(9) Click on the **Statistics** button to open the Crosstabs: Statistics dialog box (see Figure 14.4).

(10) Click on **Chi-Square**.

(11) Click on **Continue**.

(12) Click on **OK**.

The output should appear as shown in Figure 14.5.

Figure 14.4 Crosstabs: Statistics Dialog Box

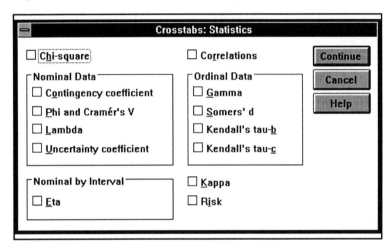

For our purposes, the only test statistic we require is the one labeled Pearson under the Chi-Square heading; this is the statistic described by textbook Eqn. 14.8. For these data the test statistic is $X^2 = 70.67$ with 4 degrees of freedom. The P value (listed under Significance) is less than .000005, leading us to conclude that role portrayal is not independent of gender. Greater proportions of males were portrayed as sellers, managers, computer experts and other, while greater proportions of females were portrayed in clerical roles.

The lowest expected frequency (15.716) is given by default. As an option, SPSS will print the expected values for each cell. To obtain the expected frequencies, follow the steps 1-12 above used for a chi-square test, and then:

(1) Click on the **Cells** button.

(2) Click on **Expected** in the Counts box.

(3) Click on **Continue**.

(4) Click on **OK**.

If many of the cells (e.g., 20% or more) have expected values below 5, the data analyst should consider combining some of the response categories. This is accomplished using the Recode procedure (Section 1.4.2) prior to conducting the chi-square test.

Figure 14.5 Chi-Square Test of Independence

```
GENDER  gender  by  ROLE  role

                    ROLE                                            Page 1 of 1
            Count
            Row Pct |seller   manager  clerical computer other
            Col Pct |                           expert                Row
                    |   1.00|    2.00|    3.00|    4.00|    5.00|  Total
GENDER      --------+--------+--------+--------+--------+--------+
              1.00  |   124  |   101  |    40  |    51  |   149  |   465
    male            |   26.7 |   21.7 |    8.6 |   11.0 |   32.0 |   70.3
                    |   60.5 |   88.6 |   46.0 |   96.2 |   73.8 |
                    +--------+--------+--------+--------+--------+
              2.00  |    81  |    13  |    47  |     2  |    53  |   196
    female          |   41.3 |    6.6 |   24.0 |    1.0 |   27.0 |   29.7
                    |   39.5 |   11.4 |   54.0 |    3.8 |   26.2 |
                    +--------+--------+--------+--------+--------+
            Column      205      114       87       53      202      661
            Total      31.0     17.2     13.2      8.0     30.6    100.0

        Chi-Square                   Value           DF         Significance
-----------------------          -----------        ----        ------------
Pearson                             70.67031          4            .00000
Likelihood Ratio                    78.02399          4            .00000
Linear-by-Linear                     5.94663          1            .01475
    Association

Minimum Expected Frequency -    15.716

Number of Missing Observations:   0
```

14.3 Measures of Association

14.3.1 The Phi Coefficient

The phi coefficient (ϕ) indicates the extent of association between two dichotomous variables. This measure of association is introduced in Chapter 6, and will be demonstrated here with two ordered variables. The data in "delinq.sav" provides information on the socioeconomic status ("SES") and population density ("pop_dens") of 75 communities in the United States. To estimate the strength of association between SES and population density, open the "delinq.sav" data file and use the following procedures:

(1) Click on **Statistics** from the menu bar.

(2) Click on **Summarize** from the pull-down menu.

(3) Click on **Crosstabs** to open the Crosstabs dialog box.

(4) Click on the row variable ("SES") and the **right arrow button**.

(5) Click on the column variable ("pop_dens") and the **right arrow button**.

(6) Click on the **Statistics** button to open the Crosstabs: Statistics dialog box.

(7) Click on **Phi and Cramer's V** in the Nominal Data box.

(8) Click on **Continue**.

(9) Click on **OK**.

The output should appear as shown in Figure 14.6.

Figure 14.6 Phi Coefficient of Association Between SES and Population Density

```
SES   by   POP_DENS   population density

                 POP_DENS       Page 1 of 1
           Count |
                 | low       high
                 |                       Row
                 |     1.00|     2.00|  Total
SES        ------+---------+---------+
             1.00|    5    |   35    |   40
      low        |         |         |  53.3
                 +---------+---------+
             2.00|   29    |    6    |   35
      high       |         |         |  46.7
                 +---------+---------+
           Column      34        41      75
            Total    45.3      54.7    100.0


                                                                Approximate
       Statistic                    Value      ASE1    Val/ASE0  Significance
    -----------------                -------   -------  -------  ------------
    Phi                             -.70508                         .00000
    Cramer's V                       .70508                          .00000

    Number of Missing Observations:  0
```

The value of the phi coefficient is approximately -0.71, indicating that the relationship between SES and population density is quite strong. By examining the cell counts, we see that low-SES communities were characterized by high population densities, while high-SES areas were primarily low population density. Because both variables are ordinal, the sign of the coefficient indicates the direction of association (low SES with high density, high SES with low density).

Section 14.3 of the textbook notes that ϕ has a simple relationship to the chi-square test of independence. The chi-square statistic is $n\phi^2 = 75 \times (-.705)^2 = 37.3$. The printed significance level (.00000) corresponds to this statistic, referring the value to the chi-square distribution with 1 degree of freedom.

14.3.2 A Coefficient Based on Prediction

Coefficient lambda (λ) is a measure of association between categorical variables based on the idea of using one variable to predict the other. Coefficient λ ranges between 0 and 1. Using the data in "exercise.sav," we shall illustrate by predicting health from exercise behavior. To use SPSS to compute this measure of association, open the data file "exercise.sav" and then:

(1) Click on **Statistics** from the menu bar.

(2) Click on **Summarize** from the pull-down menu.

(3) Click on **Crosstabs** to open the Crosstabs dialog box.

(4) Click on the row variable ("exercise") and the **right arrow button**.

(5) Click on the column variable ("health") and the **right arrow button**.

(6) Click on the **Statistics** button to open the Crosstabs: Statistics dialog box.

(7) Click on **Lambda** in the Nominal Data box.

(8) Click on **Continue**.

(9) Click on **OK**.

The output will appear as in shown Figure 14.7.

To predict health (columns) from exercise (rows) the value of the coefficient is $\lambda_{c.r} = 0.54$, which is moderate. By default, SPSS also prints $\lambda_{r.c}$ ($= 0.59$) for predicting exercise from health, and the symmetric measure λ ($= 0.57$). In this example there may be some justification for examining all three since it is plausible that healthier individuals are also more inclined to undertake vigorous exercise.

Although the textbook does not describe tests of significance for the λ-coefficients, SPSS does display P values.

Figure 14.7 Measure of Association Between Exercise and Health Using Lambda

```
EXERCISE  exercise category  by  HEALTH  health status

                     HEALTH          Page 1 of 1
              Count |
                    |good     poor
                    |
                    |   1.00|    2.00|  Row
                    |       |        |  Total
EXERCISE  ----------------------------
               1.00 |    92 |     14 |   106
   exerciser       |       |        |  52.5
                   -------------------
               2.00 |    25 |     71 |    96
non-exerciser      |       |        |  47.5
                   -------------------
             Column    117       85      202
              Total   57.9     42.1    100.0


                                                    Approximate
     Statistic              Value     ASE1    Val/ASE0  Significance
     ---------              -----     ----    --------  ------------

Lambda :
   symmetric               .56906    .06738   6.26281    .00000
   with EXERCISE dependent .59375    .06121   6.86618    .00000
   with HEALTH   dependent .54118    .07808   4.97407    .00000
Goodman & Kruskal Tau :
   with EXERCISE dependent .37764    .06739              .00000
   with HEALTH   dependent .37764    .06769              .00000

Number of Missing Observations:  0
```

14.3.3 A Coefficient Based On Ordering

Coefficient gamma (γ) is appropriate when two variables have categories that are ordered. Again, we will use the data on rates of delinquency and SES among 75 communities ("delinq.sav") since each variable is ranked as low and high. Coefficient γ is computed by SPSS using a similar procedure to that given in Section 14.3.2. Once the data file has been opened:

(1) Click on **Statistics** from the menu bar.

(2) Click on **Summarize** from the pull-down menu.

(3) Click on **Crosstabs** to open the Crosstabs dialog box.

(4) Click on the row variable ("SES") and the **right arrow button**.

(5) Click on the column variable ("delinq") and the **right arrow button**.

(6) Click on the **Statistics** button to open the Crosstabs: Statistics dialog box.

(7) Click on **Gamma** in the Ordinal Data box.

(8) Click on **Continue**.

(9) Click on **OK**.

The output should appear as shown in Figure 14.8.

Figure 14.8 Association Between Delinquency and SES Based on Gamma

```
SES    by   DELINQ   rate of juvenile delinquency
                     DELINQ           Page 1 of 1
            Count  |
                   | low      high
                   |                Row
                   |    1.00|   2.00| Total
SES         -------------------------
             1.00  |    5   |   35  |   40
  low                                   53.3
                   -------------------
             2.00  |   30   |    5  |   35
 high                                   46.7
                   -------------------
           Column       35       40       75
            Total     46.7     53.3    100.0


                                                           Approximate
       Statistic              Value     ASE1    Val/ASE0   Significance
     --------------------    -------   -------  --------   ------------

Gamma                        -.95349   .03088   -9.18013     .00000

Number of Missing Observations:   0
```

The value of gamma is -0.95, obviously indicating a very strong negative association. It is clear that high delinquency rates occur in low-SES communities and lower delinquency rates occur in high-SES communities. Although the textbook does not discuss a test of significance for γ, the SPSS output includes a P value for such a test; in the example, P is less than .000005.

Coefficient phi for the same data (output not shown) is $\phi = -0.73$. For a 2x2 table, coefficient phi also reflects the ordering of the categories; the signs of ϕ and γ will be the same. However, coefficient γ can also be used with larger tables whereas ϕ cannot.

Chapter Exercises

14.1 Using the data on restaurants ("meal.sav"), use SPSS to compute the chi-square test of independence of type of food served and cost of the meal for non-chain restaurants (Hint: use the Select If procedure).

(a) What is the value of the test statistic?
(b) Would you conclude that type of food and cost are independent?
(c) Would you draw the same conclusions for chain restaurants?

14.2 Use the data on occupations of 20 primary householders ("occup.sav") with SPSS to answer the following questions.

(a) Compute the goodness of fit statistic to test the null hypothesis that the proportions of the primary householders in occupational types are equal.
(b) Do you reject the null hypothesis at the 5% level? Circle the P value.
(c) Draw a verbal conclusion about the proportions of occupations among primary householders.

14.3 Based on the restaurant data given in "meal.sav," use SPSS to determine the extent of association between the type of food and the cost category of the meal. What type of food was most expensive in the sample? Least expensive?

14.4 Use the data on interventions aimed at reducing tobacco use among baseball players ("spit.sav"). Use SPSS to answer the following:

(a) Combine the categories for the "outcome" variable so that it is reduced to two outcomes: successful quitting and unsuccessful at quitting.
(b) Using the newly combined variable, compute the chi-square test of independence of type of intervention and outcome. Are intervention type and outcome independent?
(c) Compute the phi-coefficient and discuss whether the relationship between intervention type and outcome is strong, moderate, or weak.
(d) Compute $n\phi^2$ to verify that it is equal to the value of the chi-square test statistic.

14.5 Use the data on voting behaviors in "vote.sav," use SPSS to answer the following:

(a) Would you support the hypothesis that voting behavior in the last election is independent of political party?
(b) What is the value of the test statistic that supports this conclusion?
(c) Would you use the gamma statistic to indicate the extent of association between previous voting behavior and political party affiliation? If not, which measure of association would be appropriate?

14.6 Create a new data file with the data of Table 14.31 in the textbook. From this file:

(a) Use SPSS to test whether individuals chosen for portrayal in computer magazines represent a random sample of American men, women, boys, and girls. (Hint: you will need to use the hypothesized proportions from Table 14.31 also).

(b) What are the values of the test statistic and P for the goodness-of-fit test?
(c) Draw a conclusion about the distribution of portrayals on computer magazines.

Appendix XIV
SPSS Syntax for Inference on Categorical Data

14.1 Tests of Goodness of Fit

To test the goodness of fit for one categorical variable ("sex") using the chi-square statistic, use the following command:

Npar tests chisquare = sex (1,2).

14.2 Chi-Square Tests of Independence

To compute the chi-square test of independence for "gender" and "role" portrayal, use the following SPSS syntax:

Crosstabs tables = gender by role
   /statistics=chisq.

14.3.1 The Phi Coefficient

To compute the phi coefficient of association between "SES" and population density ("pop_dens"), use the following command:

Crosstabs tables = ses by pop_dens
   /statistics=phi.

14.3.2 A Coefficient Based on Prediction

To compute the lambda coeffient for predicting future voting behavior ("voting") from past voting behavior ("voted"), the SPSS syntax is:

Crosstabs tables = voted by voting
   /statistics=lambda.

14.3.3 A Coefficient Based on Ordering

To compute the Gamma coeffient for ordered categorical variables, use the following command:

Crosstabs tables = ses by delinq
   /statistics=gamma.

Chapter 15   SIMPLE REGRESSION ANALYSIS

This chapter describes how to use SPSS for Windows to perform simple linear regression analysis. Because a scatterplot and correlation coefficient are indispensible in interpreting regression results, procedures for obtaining these are reviewed as well. The SPSS output for simple regression analysis includes many results that are redundant. This is because the same regression output pertains to multiple regression analysis as well, in which the results tell us about different aspects of the data. Multiple regression analysis (regression with two or more independent variables) is not discussed in the textbook and so is not demonstrated in this manual.

15.1 The Scatter Plot and Correlation Coefficient

An important step in regression analysis is to examine a scatter plot of the two variables and to calculate the correlation coefficient. Although both of these procedures are described in Chapter 5, we will illustrate them here using the "cancer.sav" data file. In this example, we wish to examine the relationship between the amount of exposure to radioactive materials and the cancer mortality rate. To create a scatter plot of these variables, open the data file and:

(1) Click on **Graphs** from the menu bar.

(2) Click on **Scatter** to open the Scatter dialog box.

(3) Click on **Define** to open the Simple Scatterplot dialog box.

(4) Click on the independent variable (x) that you wish to examine ("expose") and the **right arrow button** to move it into the X Axis box.

(5) Click on the dependent variable (y) that you wish to examine ("mortalit") and the **top right arrow button** to move it into the Y Axis box.

(6) Click on **OK**.

The correlation coefficient can be calculated by using the following commands:

(1) Click on **Statistics** from the menu bar.

(2) Click on **Correlate** from the pull-down menu.

(3) Click on **Bivariate** to open the Bivariate Correlations dialog box.

(4) Click on the variable(s) that you wish to correlate, each followed by the **right arrow button** to move them into the variables box.

(5) Make sure that the Display actual significance level box is checked. Click on **OK**.

The output for the scatter plot and correlation of exposure index and cancer mortality appears in Figure 15.1.

Figure 15.1 Scatter Plot and Correlation Coefficient

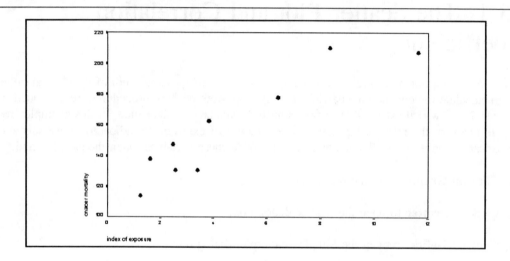

```
                       - -  Correlation Coefficients   - -

                  EXPOSE        MORTALIT

EXPOSE           1.0000           .9263
                 (      9)        (      9)
                 P=  .            P=  .000


MORTALIT          .9263          1.0000
                 (      9)        (      9)
                 P=  .000         P=  .
(Coefficient / (Cases) / 2-tailed Significance)
```

The swarm of points in the scatter plot goes from lower left to upper right. We also see that there are no apparent outliers. The correlation between exposure and mortality is +0.93, indicating that it is both positive and strong. Thus, higher levels of exposure to radioactive materials are strongly associated with higher levels of cancer mortality. The P value listed directly under the correlation results from a test of significance (t-test of the hypothesis that the correlation is zero); this is discussed in the section "Test of Significance for the Correlation."

15.2 SPSS for Simple Regression Analysis

All of the statistics needed for simple regression analysis are obtained in a single procedure. We will illustrate this procedure using the "cancer.sav" data set to examine the association between mortality rates and exposure to radioactive materials.

After opening the "cancer.sav" data file, the steps for a regression analysis are:

(1) Click on **Statistics** on the menu bar.

(2) Click on **Regression** from the pull-down menu.

(3) Click on **Linear** to open the Regression dialog box (see Figure 15.2).

(4) Click on the variable that is your dependent variable ("mortalit"), and then click on the **right arrow button** to move the variable name into the dependent variable box.

(5) Click on the variable that is your independent variable ("expose"), and then click on the **right arrow button** to move the variable name into the independent(s) variable box.

(6) Click on **OK**.

Figure 15.2 Regression Dialog Box

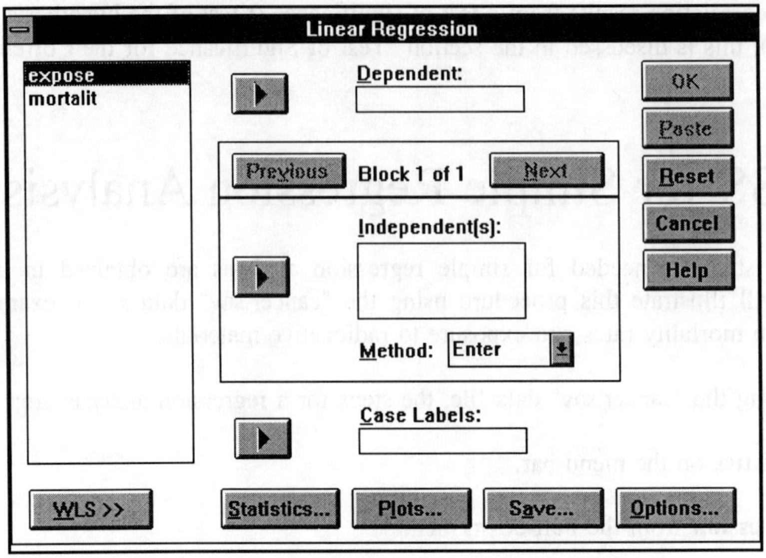

[Linear Regression dialog box screenshot]

One optional result is useful as well, namely, a confidence interval for the slope of the regression line. To obtain a 95% interval, follow steps 1-5 above and then:

(1) Click on **Statistics** to open the Regression: Statistics dialog box.

(2) Click on **Confidence Intervals** in the Regression Coefficients box.

(3) Click on **Continue**.

(4) Click on **OK**.

The complete output is shown in Figure 15.3.

Figure 15.3 Regression Analysis Output with Confidence Interval for Beta

```
                * * * *    M U L T I P L E   R E G R E S S I O N   * * * *

Listwise Deletion of Missing Data

Equation Number 1    Dependent Variable..    MORTALIT    cancer mortality

Block Number  1.  Method:  Enter        EXPOSE

Variable(s) Entered on Step Number
   1..    EXPOSE      index of exposure

Multiple R              .92634
R Square                .85811
Adjusted R Square       .83785
Standard Error        14.00993

Analysis of Variance
                    DF       Sum of Squares      Mean Square
Regression           1          8309.55586       8309.55586
Residual             7          1373.94636        196.27805

F =     42.33563       Signif F =   .0003

------------------- Variables in the Equation ----------------------
Variable              B         SE B      95% Confdnce Intrvl B       Beta

EXPOSE           9.231456    1.418787     5.876566    12.586347      .926345
(Constant)     114.715631    8.045663    95.690703   133.740559


----------- in ------------

Variable            T     Sig T

EXPOSE           6.507   .0003
(Constant)      14.258   .0000


End Block Number   1   All requested variables entered.
```

Estimating the Regression Equation

The least-squares estimates of the intercept and slope of the regression line are displayed in the center portion of the output (Figure 15.3) under the title "Variables in the Equation." Two values are listed in the column headed "B"; these are the regression weight (9.23) and the intercept (114.72), respectively, computed to several decimal places. The equation of the least-squares line is thus y = 114.72 + 9.23x. (Instructions for having SPSS add the regression line to the scatter plot are given in a later section.)

SPSS also prints a form of β called the "standardized regression weight," labeled Beta in the output. Although not discussed in the textbook, the standardized weight is the *number of standard deviations change in y associated with a one-standard deviation change in x*. Thus, in this example, a one-standard deviation increment in exposure is associated with a 0.93-standard deviation increment in cancer mortality -- a large effect. When the units of x and y are familiar -- for example, income, waiting time, body weight -- the unstandardized ("raw") coefficient is easily interpreted. When scales are in less familiar units -- for example, psychological test scores -- the standardized weight is a convenient way to express the relationship of x with y.

The Confidence Interval option produced two 95% intervals in the output, one for the slope and one for the intercept. The interval for the slope indicates that we are 95% confident that a one-unit increment in exposure is associated with an increase in mortality of at least 5.88 deaths per 100,000 person years and perhaps as much as 12.59 additional deaths. These values were obtained by adding to and subtracting from 9.23 a multiple of the standard error required to give the preselected confidence interval. In this example, the standard error[1] is 1.42 (see SE B in Figure 15.3), and the multiplier from the t-distribution with 7 degrees of freedom is 2.37.

Test of Significance for β

A test of the hypotheses H_0: $\beta = 0$ and H_1: $\beta \neq 0$ is given at the bottom of the regression output (Figure 15.3). The t-statistic is t = 9.23/1.418 = 6.507. The P value, listed under Sig T in the output, is .0003. Since this is smaller than any reasonable value of α (e.g., .05 or .01 or even .001), H_0 is rejected; there is a nonzero (positive) association between exposure and cancer mortality in the population of counties represented by this sample.

At the outset of this study, researchers had reason to believe that a positive association might be found. Thus, a one-tailed test would have been appropriate with H_0: $\beta \leq 0$ and H_1: $\beta > 0$. The P value printed by SPSS is for a two-tailed test. To reject H_0 in favor of a one-sided alternative, P/2 must be less than α and the sign of the regression weight must be consistent with H_1. Both conditions are met in this example and H_0 is rejected.

---

[1]The 1996 printing of the Anderson-Finn book has an error in line 18 on page 580. The standard error used in the confidence interval was not correct, that is, 0.467 was used instead of 1.419 given in line 8. The correct confidence interval is (4.26, 14.20).

Strength of Association of x and y

The correlation coefficient provides a numerical index of the strength of association between the independent and dependent variable.[2] The regression procedure in SPSS does not compute the correlation directly, and thus it is advisable to refer to the correlation output (Figure 15.1). For this example, the value 0.926 is both positive and large relative to the maximum value of 1.

The square of the correlation ($0.926^2 = 0.858$) is the proportion of variation in y attributable to x; that is, 85.8% of the variation in cancer mortality is attributable to radiation exposure. Obviously this is a very strong association.

These statistics are available indirectly in the regression output (Figure 15.3) because of the redundancy in regression analysis with one predictor variable. First, the top portion of the output lists R Square. Although this is the square of the statistic called the "multiple correlation," in simple regression it is equivalent to the square of the correlation coefficient, that is, $r^2 = 0.858$. The result labeled in the output as Multiple R is the multiple correlation itself. Its absolute value is the same as r, but unlike the correlations discussed in the textbook, the multiple correlation is always positive. The data analyst must remember that the multiple correlation does not indicate the direction of association! Second, if you studied the output in detail you may have noticed that the standardized regression weight (Beta = 0.926) is also equal to the correlation. These two statistics are algebraically identical in regression analysis with one independent variable (x); they are not equivalent when the analysis has more than one x variable.

The output also includes a table labeled Analysis of Variance located below the information about multiple correlation. The sum of squares labeled Regression (8309.56) corresponds to the numerator of textbook Equation 15.7. The sum of squares labeled Residual (1373.95) is the sum of squared differences between the predicted values and the actual values of y, that is, the sum of squared deviations of the data around the regression line. These are combined according to Equation 15.7 to yield the proportion-of-variation statistic, r^2. The residual sum of squares divided by the number of degrees of freedom (7) is the variance of the residuals, 196.28 in the column labeled Mean Squares. The square root of this value, 14.01, is the standard error of estimate $s_{y \cdot x}$.

Test of Significance for the Correlation

The correlation output (Figure 15.1) also includes a test of the correlation coefficient with hypotheses $H_0: \rho = 0$ and $H_1: \rho \neq 0$. The t-statistic (not printed) is

---

[2]The results in the textbook are based on summary statistics that have been rounded and thus do not agree exactly with those in Figure 15.3.

$t = 0.926 \times \sqrt{(9-2)/(1-0.926^2)} = 6.507$. Instead, a P value is given, obtained by referring this statistic to the t-distribution with 7 degrees of freedom. Since P is very small, H_0 is rejected at any reasonable significance level (α). If a one-tailed test is required, P/2 must be less than α and the sample correlation must be in the direction specified by H_1 in order for H_0 to be rejected.

The careful reader may notice that this t-value and P are the same as those for testing the significance of the regression weight. When a study has just one numerical independent variable and one numerical dependent variable, the regression coefficient and the correlation coefficient have the same sign (+ or -) and the tests of significance are identical.

Drawing the Regression Line

As a separate procedure, SPSS will draw the least-squares line (regression line). Once you have created the scatter plot, use the following instructions to draw the regression line:

(1) With the Scatter Plot opened (Chart Carousel Window), click the **Edit** button.

(2) Click on **Chart** in the menu bar.

(3) Click on **Options** from the pull-down menu to open the Scatterplot: Options dialog box.

(4) In the Fit Line box, Click on **Total** and then click on the **Fit Options button**.

(5) Click on **Continue**.

(6) Click on **OK**.

The output is shown in Figure 15.4.

Notice that this is the same plot as in Figure 15.1 with the least-squares line superimposed. (You may wish to compare this with textbook Figure 15.5.) If there were any outliers in the data set, they would be conspicuous by their (vertical) distances from the regression line.

Figure 15.4 Scatter Plot with Regression Line

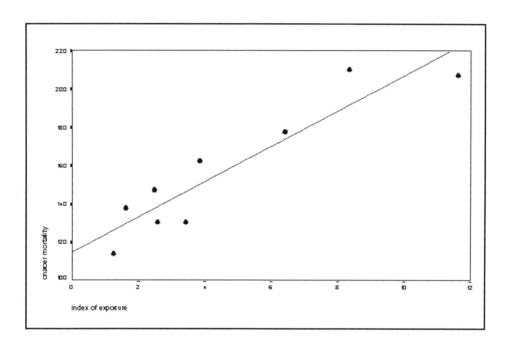

15.3 Another Example: Inverse Association of x and y

As another example, open the data file "noise.sav" which has data on the relationship between the acceleration noise of an automobile and the speed for a section of highway. Make a scatter plot (with the regression line superimposed) with acceleration noise as the dependent variable (y) and speed (mph) as the independent variable (x); compute the correlation coefficient and perform a simple regression analysis using the steps outlined in Sections 15.1 and 15.2. Your output should look like Figure 15.5.

Figure 15.5 Scatter Plot, Correlation, and Regression Analysis of Noise and Speed

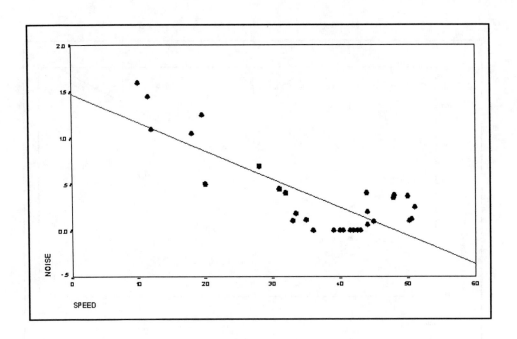

```
                - -   Correlation Coefficients   - -

                  NOISE         SPEED

NOISE            1.0000         -.8178
               (    30)        (    30)
               P= .            P= .000

SPEED           -.8178         1.0000
               (    30)        (    30)
               P= .000         P= .
```

(Coefficient / (Cases) / 2-tailed Significance)

" . " is printed if a coefficient cannot be computed

(Continued)

```
* * * *   M U L T I P L E     R E G R E S S I O N   * *
```
Listwise Deletion of Missing Data

Equation Number 1   Dependent Variable..   NOISE

Block Number   1. Method: Enter   SPEED

Variable(s) Entered on Step Number
    1..   SPEED

Multiple R	.81777
R Square	.66875
Adjusted R Square	.65692
Standard Error	.27100

Analysis of Variance

	DF	Sum of Squares	Mean Square
Regression	1	4.15168	4.15168
Residual	28	2.05642	.07344

F =   56.52877   Signif F =   .0000

------------------ Variables in the Equation ------------------

Variable	B	SE B	Beta	T	Sig T
SPEED	-.030649	.004076	-.817772	-7.519	.0000
(Constant)	1.479673	.155202		9.534	.0000

End Block Number   1   All requested variables entered.

Notice that there is an inverse relationship between speed and noise. We first see this with the negative correlation (−0.818), indicating a strong, negative linear association. The inverse relationship is also indicated by the negative slope of the line in the scatter plot. The scatter plot also shows that there are no apparent outliers.

The test of whether the regression weight is significantly different from zero appears at the lower right in the regression output. The sample regression weight is −0.031. The t-statistic is t = −0.0306/0.0041 = −7.519 where 0.0041 is the standard error of the regression coefficient. The P value given under the label Sig T is .0000 which implies that P < .00005; noise is significantly inversely related to speed on the highway.

A one-tailed test would have been reasonable in this study with the hypothesis $H_0: \beta \geq 0$ and $H_1: \beta < 0$. To reject H_0, P/2 must be less than α and the sign of the sample regression weight must be consistent with H_1 (that is, negative). Both conditions are met by these data, assuming any reasonable significance level (e.g., .05, .01, or .001).

Given that we have a significant negative association of noise with speed, we will ask about the strength of the relationship. Since speed is measured in familiar units (miles per hour, or mph), we may prefer to interpret the unstandardized regression weight (labeled B in the output). This tells us that every 1 mph increase in average speed on sections of the highway is associated with a .031-unit decrease in acceleration noise.

We have already seen that the correlation between speed and noise is strong and negative. In addition, the output produced by the regression analysis shows that the square of the correlation is 0.669. (Recall that the Multiple R is the absolute value of the correlation.) Thus, 66.9% of the variability in noise level is accounted for by speed and, by subtraction from 100%, 33.1% of variation in noise level is explained by other factors that were not included in this study.

No Relationship

Does it ever happen that a predictor variable is not related to the dependent variable? The answer is yes; this can be illustrated with the data in file "weather.sav" by examining the relationship between the amount of precipitation in an area and the temperature. Again, construct a scatter plot of the variables amount of precipitation ("precip") and temperature ("temp"), compute the correlation and perform a regression analysis using the steps listed in Sections 15.1 and 15.2. The output is displayed in Figure 15.6.

It is extremely difficult to draw a single straight line through the swarm of points in the scatter plot. In fact, the regression line appears to be rather flat and many of the points are far from it. It is also difficult to tell whether the swarm of points forms a pattern that runs from lower left to upper right or from upper left to lower right. The correlation itself is small (r = 0.124) and nonsignificant (P = .284); less than 2% of the variability in "precipitation" is attributed to "temperature" (0.124^2 = 0.015). There is little to be gained by attempting to predict precipitation from temperature.

Figure 15.6 Scatter Plot and Regression Analysis of Precipitation and Weather

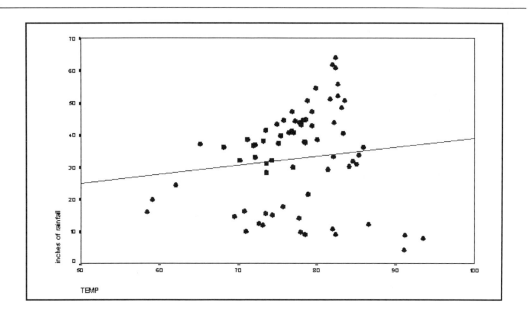

```
        - -  Correlation Coefficients  - -

                    PRECIP        TEMP

     PRECIP        1.0000        .1237
                  (    77)      (    77)
                   P=  .         P=  .284

     TEMP           .1237       1.0000
                  (    77)      (    77)
                   P=  .284      P=  .


(Coefficient / (Cases) / 2-tailed Significance)

" . " is printed if a coefficient cannot be computed
```

Chapter Exercises

15.1 Enter the data in Table 15.8 in the textbook into a new file.

(a) Use SPSS to make a scatter plot of the variables "volumes" and "staff," and draw the least-squares line through the scatter plot.
(b) Is the relationship between libraries' collection size and staff size positive or negative?
(c) Judging from the scatter plot, would you estimate that the correlation is weak, moderate, or strong?

15.2 Enter the data on competitiveness and productivity in Table 5.13 in the text and use SPSS to:

(a) Perform a regression analysis to examine whether one's productivity is related to how competitive he/she is.
(b) Compute the slope, the standard error of the slope, and a 95% confidence interval for the slope of the regression line.
(c) Is there a significant relationship between competitiveness and productivity? Label the test statistic and P value in the output.
(d) What is the value of the correlation between competitiveness and productivity?
(e) Circle the value that is the square of the correlation, and write a sentence of interpretation for this result.

15.3 Open the "fire.sav" data file and use the variables "sex" and "agility" to perform the following:

(a) Use SPSS to recode the sex variable so that males have a value of 1 and females have a value of 0. (Note: this is called "dummy coding" a variable.)
(b) Using SPSS, perform a regression analysis with the agility score as the dependent variable and sex, after recoding, as the independent variable.
(c) Is there a statistically significant relationship of agility with sex? What is the direction of the relationship? Be clear about which "sex" code is associated with better (lower) agility scores.
(d) By hand or with SPSS, calculate the mean agility scores for males and for females and the difference between the two means. Compare this difference with the regression weight you obtained.
(e) Given the choice between the raw and standardized regression coefficient, which would you choose to emphasize in a written report of this study? Why?

Appendix XV
SPSS Syntax for Simple Regression Analysis

15.1 The Scatter Plot and Correlation Coefficient

To create a scatter plot of exposure index and mortality rate, use the following command:

Plot plot = mortalit with expose.

15.2 Simple Regression Analysis

To run a regression analysis along with a 95% confidence interval for the unstandardized regression coefficient:

Regression
  /statistics=coeff outs ci r anova
  /dependent=mortalit
  /method=enter expose.

Chapter 16 COMPARISONS OF SEVERAL POPULATIONS

This chapter describes the analysis of variance used to compare the means of two or more populations. The SPSS procedure One-Way is used to compute the ANOVA table and perform the hypothesis test. The methods employed by SPSS for calculating follow-up tests for specific mean differences and for estimating effect sizes are different than those discussed in your textbook, so we will compute these by hand.

16.1 One-Way Analysis of Variance

16.1.1 A Complete Example

We illustrate the ANOVA procedure using SPSS with the "words.sav" data file, which contains the data from Table 16.1 in the textbook. This file has two variables -- one indicating the information set provided to the child (1=no information, 2=three categories, 3=six categories) and the other being the number of words memorized by the child. We shall use SPSS to test that the mean number of words memorized by the three groups are equal, that is, $H_0: \mu_1 = \mu_2 = \mu_3$.

Examining the Data

It is always important to examine the data visually prior to conducting such a test. While SPSS does not make line graphs like Figure 16.2 in the textbook, a scatter plot of the data is an acceptable alternative. The procedure for creating scatter plots is introduced in Chapter 5 but several additions may be necessary. First, you may need to change the increment size of the y-axis in order to distinguish close values of the outcome variable. Second, you may wish to have "duplicate" values indicated clearly in the plot.

Figure 16.1 displays the graph of the number of words memorized by the three different information sets. To obtain this graph, the default chart produced by SPSS would be edited as follows:

To change the increment on the y-axis to 1:

(1) From the Chart Carousel window, click on **Edit** to open the Chart window.

(2) Click on **Chart** from the main menu.

(3) Click on **Axis** from the pull-down menu to open the Axis Selection dialog box.

(4) Click on **Y Scale** and then click on **OK**.

(5) In the Major Divisions part of the Y Scale Axis dialog box, change the "Increment" from 2 to 1 and click on **OK**.

To display duplicate points using the "sunflower" option:

(1) Click on **Chart** from the main menu.

(2) Click on **Options** from the pull-down menu.

(3) In the Sunflowers part of the Scatterplot Options dialog box, click on **Show Sunflowers** and then click on **OK** to redraw the graph.

Figure 16.1 Scatter Plot with Sunflowers Showing Multiple Data Points

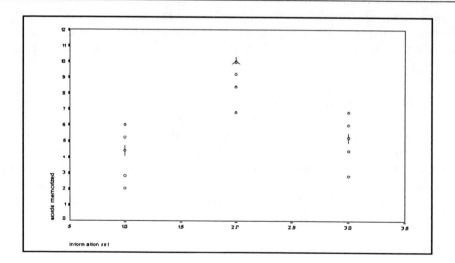

In Figure 16.1, the open circle represents one data point. For data points with "petals" (lines running through the circle), the number of petals indicates the frequency of observations at that specific point. Note that there are two children in the "no information" group who memorized 4 words, and three children in the "three categories" group who memorized 10 words.

Running the One-Way Procedure

To direct SPSS to perform the One-Way procedure:

(1) Click on **Statistics** from the menu bar.

(2) Click on **Compare Means** from the pull-down menu.

(3) Click on **One-Way ANOVA** from the pull-down menu to open the One-Way ANOVA dialog box (Figure 16.2).

(4) Click on and move the "words" variable to the Dependent List box using the **upper right arrow button**.

(5) Click on and move the "info_set" variable to the Factor box using the **lower right arrow button**.

(6) Click on the **Define Range button** to open the One-Way ANOVA: Define Range dialog box.

Figure 16.2 One-Way ANOVA Dialog Box

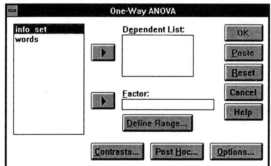

(7) The range of values for the "info_set" factor is 1 to 3. Enter the value 1 in the Minimum box, and 3 in the Maximum box.

(8) Click on **Continue** to close the dialog box.

(9) Click on the **Options button** in the lower right corner of the One-Way ANOVA dialog box.

(10) Click on the **Descriptives** option of the Statistics box. (This option provides mean, standard deviation, confidence interval, standard error, and the minimum and maximum for each information category separately.)

(11) Click on the **Display Labels** option.

(12) Click on **Continue** to close the dialog box.

(13) Click on **OK** to run the procedure.

The ANOVA output is displayed in Figure 16.3. The analysis of variance table in this listing is similar to Table 16.5 of the textbook. The F Prob. (.0000) represents the P value corresponding to the F-Ratio of 22.5 with 2 and 15 degrees of freedom. The null hypothesis that the population means are equal is rejected for any α greater than or equal to .00005. Thus, we conclude that there are differences among the three groups in the mean numbers of words memorized based on the information set.

The middle and lower portions of this listing is the result of our requesting descriptive statistics through the Descriptives option. From this, we see that the average number of words remembered by the no information group was 4; by the three categories group it was 9; and by the six categories group it was 5. SPSS also lists the range of words remembered for each group and computes a 95% confidence interval for each of the means. The row labeled Total gives the same results for the entire sample of 18 observations.

Figure 16.3 One-way ANOVA Listing

```
- - - - - O N E W A Y - - - - -
      Variable   WORDS      words memorized
   By Variable   INFO_SET   information set

                          Analysis of Variance

                              Sum of            Mean              F         F
           Source      D.F.   Squares           Squares            Ratio     Prob.

Between Groups          2     84.0000           42.0000            22.5000   .0000
Within Groups          15     28.0000            1.8667
Total                  17    112.0000


                              Standard  Standard
Group         Count    Mean   Deviation  Error    95 Pct Conf Int for Mean
no infor        6     4.0000  1.4142    .5774      2.5159 TO    5.4841
3 catego        6     9.0000  1.2649    .5164      7.6726 TO   10.3274
6 catego        6     5.0000  1.4142    .5774      3.5159 TO    6.4841

Total          18     6.0000  2.5668    .6050      4.7236 TO    7.2764

GROUP         MINIMUM  MAXIMUM
no infor      2.0000    6.0000
3 catego      7.0000   10.0000
6 catego      3.0000    7.0000

TOTAL         2.0000   10.0000
```

16.1.2 An Example with Unequal Sample Sizes

The method for performing analysis of variance with unequal cell sizes with SPSS is the same as that for equal sample sizes. Using the "final.sav" data file and the procedure in Section 16.1.1 above, we can perform an ANOVA to test the hypothesis that there is no difference in average course grade between native English speaking students and other students. We would follow steps 1-13 of Section 16.1.1 substituting the factor "language" with two levels for the factor "info_set," and the variable "grade" for "words." The results are listed in Figure 16.4.

This output is identical in format to Figure 16.3. Notice that there are 15 students for whom English is not their first language, but 53 native English speakers in the sample. The F-ratio is large (7.4715), and the P value is .0080. From this, we conclude that there are differences in final course grade based on language. In fact, because there are only two groups to compare in this analysis, we can conclude that native English speakers perform better, on average. The sample mean for native English speakers was 136.00 and for non-native speakers it was 119.73.

Figure 16.4 One-way ANOVA Output for Two Groups of Unequal Size

```
             - - - - -  O N E W A Y  - - - - -
     Variable   GRADE
  By Variable   LANGUAGE    native language

                        Analysis of Variance

                            Sum of        Mean            F       F
       Source       D.F.   Squares       Squares        Ratio   Prob.

Between Groups        1    3093.5373    3093.5373      7.4715   .0080
Within Groups        66   27326.9333     414.0444
Total                67   30420.4706


                              Standard   Standard
Group      Count      Mean    Deviation   Error    95 Pct Conf Int for Mean

other        15    119.7333   23.1099    5.9669    106.9355  TO   132.5312
English      53    136.0000   19.5379    2.6837    130.6147  TO   141.3853

Total        68    132.4118   21.3081    2.5840    127.2541  TO   137.5694

GROUP      MINIMUM     MAXIMUM

other      83.0000    157.0000
English    85.0000    165.0000

TOTAL      83.0000    165.0000
```

16.1.3 More About the Analysis of Variance

As with other statistical procedures, it is important to evaluate whether the data meet the assumptions underlying the ANOVA. These assumptions include (1) independence of observations, (2) a normal distribution of subgroup means, and (3) homogeneity of population variances.

While the analysis of variance is fairly robust with regard to violation of conditions (2) and (3), these assumptions should be examined routinely, especially if sample sizes are small to moderate. You can, for instance, inspect histograms of the dependent variable to assess departures from normality, as illustrated in Section 11.1.2. Chapter 13 gives a formal test for determining whether or not variances of two groups are equal. When there are more than two populations involved, the sample variances should be examined to see if they are similar. In Figure 16.3, for example, the standard deviation of the three groups are in the same general range.

16.2 Which Groups Differ from Which, and by How Much?

16.2.1 Comparing Two Means

When population means differ, it is informative to estimate the degree of difference. If the scale of the variable is well known, then the raw mean difference may be a meaningful way to describe the magnitude of the difference. If the scale is not familiar, estimating the magnitude of a mean difference is often accomplished by computing an effect size, which is the mean difference expressed in standard deviation units.

SPSS does not calculate effect sizes in the One-Way ANOVA procedure, so they must be computed by hand. In most studies, the appropriate standard deviation is the common within- cell standard deviation; this is the square root of the mean-square within groups. In the final statistics grade example (Figure 16.3), the effect size for the difference between native and non-native English speakers is $(136.00 - 119.73)/\sqrt{414.04} = 0.80$. On average, native English speakers received grades in statistics that were 0.80 standard deviations above non-native English speakers.

We can also compute a confidence interval for the difference of two means using the approach given in Section 16.2 of the textbook. In the final statistics grade example, the t-value for a 95% confidence interval is $t_{66}(.025)$. From Appendix III in the textbook, the t-value is approximately equal to 2.000. The mean difference is $136.00 - 119.73 = 16.27$ and the standard error of the differences is $\sqrt{414.04 \times (1/15 + 1/53)} = 5.95$. The lower bound of the confidence interval is 16.27 − (2.00 x 5.95) = 4.37, and the upper bound is 16.27 + (2.00 x 5.95) = 27.17. The 95% confidence interval for the difference between average final grades between native and non-native English speakers is (4.37, 28.17). All plausible values for the mean difference favor native speakers.

16.2.2 More Than Two Means

Tests of Specific Differences

When there are more than two populations involved in analysis of variance, it is impossible to tell from the overall F-ratio which means differ significantly from each other means simply by inspection. You must conduct individual tests for the particular pairs of means you wish to compare. Conducting multiple tests inflates the probability of a type I error, however, so that some type of adjustment, such as the Bonferroni procedure, is necessary.

The SPSS ANOVA procedure does have an option for computing Bonferroni comparisons, but it uses a "modified" approach; therefore, we will calculate the tests by hand. The standard ANOVA printout contains all of the values needed to compute the t-statistic. In the three-group example (Figure 16.3) there are three possible pairwise comparisons that may be made. For the test of the "no information" compared to "three categories" group, the test statistic is t = (4 − 9) / [$\sqrt{1.87}$ x (1/6 + 1/6)] = −6.34. Similarly, the t-statistic for six categories compared to three categories is t = −5.07, and for six categories compared to no information is t = 1.27.

If we choose a familywise α-level of .03, the α for individual comparisons is $\alpha^* $ = .03/3 = .01. The significance points for a two-tailed test from the t-distribution with 15 degrees of freedom are ±2.947. Therefore, we conclude that the three-category method is superior to both the no information and the six-category method. There is no difference in average performance between the no information and six-category approaches, however.

Effect Sizes

The t-test allows us to determine which means differ from which other means, but does not provide a clear indication of the magnitude of these differences. In some cases, the difference itself is a meaningful way to describe the strength of the effect. For example, students in the three categories condition remember, on average, 9 − 4 = 5 more words than those in the no information condition.

A confidence interval for the difference may also be obtained as in the example in Section 16.2.1. Here we would use the Bonferroni α^* as the confidence coefficient.

There are other instances when the scale of the dependent variable is not familiar, as with many psychological tests. In such a situation, an effect size may be more meaningful. For example, the 5-word advantage of three-categories over no information corresponds to an effect size of $5/\sqrt{1.87} = 3.66$ standard deviations, a large effect indeed.

16.3 Analysis of Variance of Ranks

When the assumptions of normality and/or homogeneity of variance are severely violated, an alternative procedure is the Kruskal-Wallis analysis of variance of ranks. The data file "bottle.sav" contains the daily output for three bottle capping machines, as listed in Table 16.11 of the textbook. Using this data file, the procedure for using SPSS to perform the analysis of variance of ranks is:

(1) Click on **Statistics** from the menu bar.

(2) Click on **Nonparametric Tests** from the pull-down menu.

(3) Click on **K independent samples** from the pull-down menu. This opens the Tests for Several Independent Samples dialog box (Figure 16.5).

Figure 16.5 Test for Several Independent Samples Dialog Box

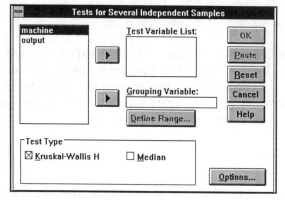

(4) Click on and move the "output" variable to the Test Variable List box using the **upper right arrow button**.

(5) Click on and move the "machine" variable to the Grouping Variable box using the **lower right arrow button**.

(6) Click on the **Define Range button** to open the Several Independent Samples: Define Range

dialog box.

(7) The machine variable is coded 1 through 3, so enter **1** in the minimum box and **3** in the maximum box.

(8) Click on **Continue** to close the dialog box.

(9) Notice that the Kruskal-Wallis H test is the default option in the Test Type box. Therefore, click on **OK** to run the procedure.

The output from this test in contained in Figure 16.6. It shows the mean rank and sample size for each machine. For example, there are 4 machines of type C, and their mean rank is 10.00. The listing also reports the chi-square statistic (5.66) and the P value (.0591). Using an α level of .05, the null hypothesis is accepted and we conclude that the machines do not differ with respect to bottle cap output.

Figure 16.6 Kruskal-Wallis One-Way ANOVA Listing

```
- - - - - Kruskal-Wallis 1-Way Anova

     OUTPUT      bottle cap output
  by MACHINE     machine

    Mean Rank      Cases

         4.80         5    MACHINE =   1    Machnie A
         4.67         3    MACHINE =   2    Machine B
        10.00         4    MACHINE =   3    Machine C

                      --

                      12    Total

   Chi-Square      D.F.    Significance
       5.6564         2            .0591
```

Chapter Exercises

16.1 Using the semesters of statistics and grade variables of the "semester.sav" data file do the following:

(a) Use SPSS to recode the "semester" variable into three categories -- no statistics, one semester, and more than one semester.
(b) Test the hypothesis that there are no differences, on average, in test scores based on prior experience in statistics. Use $\alpha=.05$ and state your conclusions in words.
(c) Using the Bonferroni adjustment and a familywise α level of .06, conduct all pairwise comparisons to determine which groups differ from one another. State your conclusions in one or two sentences.
(d) Compute an effect size for any of the comparisons that were significant in (c).

16.2 The data file "movies.sav" contains information regarding yearly gross sales for the top 100 films from 1995. The list was compiled by *Daily Variety*. Box office figures include revenues from the United States and Canada, and are reported in U.S. dollars. The movies are also categorized by genre (1=action-adventure, 2=drama, 3=children's, 4=comedy).

Assume that you are working at a small production company that can only produce one film a year. You wonder of it would be beneficial to you to consider one type of film over another, and plan to use SPSS to conduct an ANOVA on gross sales by genre.

(a) Before running the ANOVA procedure, use SPSS to find the variances of gross amount by genre. Do they appear equal? What is the ratio of the largest variance to the smallest variance?
(b) Examine the histogram of the "gross" variable. What is the shape of the distribution?
(c) Compute a new variable, "lngross" which is the natural log of the gross variable. (Hint: Use the Compute procedure and enter LN(gross) in the Numeric Expression box.)
(d) Repeat steps (a) and (b) using the "lngross" variable. Does this transformed variable seem to better satisfy the assumptions of ANOVA (e.g., has the ratio of largest to smallest variance been reduced?)
(e) Use SPSS to perform the ANOVA. What are your recommendations regarding concentrating on a specific genre of film to produce?

16.3 The "phone.sav" data file contains the number of monthly telephone calls to an airline reservations desk for three consecutive years, as listed in Table 4.8 of the textbook. Using these data, answer the following:

(a) Use SPSS to conduct an ANOVA to test whether the mean number of calls is the same from year to year. What is your conclusion if $\alpha = .05$? Indicate the range of α for which you would reject the null hypothesis (i.e., the P value).
(b) Make all pairwise comparisons to determine what years differ from one another. (Chose an appropriate α level, and state the significance points, the value of each test statistic, and the conclusion for each comparison.)
(c) Report the nature (magnitude and direction) of any significant differences you found in part (b) in the most easily interpretable metric. Justify your choice.
(d) Repeat the ANOVA, this time using the ranks (Kruskal-Wallis) procedure. How does your conclusion from this method differs from that of part (a) (if at all)?
(e) Evaluate the validity of the assumptions of normality and equal variances, and indicate which method of analysis is preferred based on these results.

16.4 The "salary.sav" file contains the data on salaries for a sample of men and women contained in Table 12.8 of the textbook. Use it to do the following:

(a) Conduct a t-test to examine potential differences in average salary between males and females. Do salaries differ, on average? Which group has a higher salary?
(b) Repeat the test in part (a) using the one-way ANOVA procedure. Are your results the same?

Appendix XVI
SPSS Syntax for Comparisons of Several Populations

16.1.1 A Complete Example

Use the following command to perform a one-way analysis of variance of the "words" variable by the levels of the "info_set" variable:

Oneway words by info_set(1,3)
/statistics = descriptives.

16.3 Analysis of Variance of Ranks

Use the following syntax to perform the Kruskal-Wallis analysis of variance of ranks for the variable "output" by the variable "machine:"

Npar tests k-w = output by machine(1,3).

Appendix: DATA FILES

All of the following data sets are located at the Springer Internet address given in the Preface. Each data set contains a ".sav" and ".dat" file. The ".sav" files are SPSS data files, and the ".dat" files are ASCII files.

BOTTLE.DAT Daily Output of 12 Bottle Capping Machines (Table 16.11)

Variable	Column(s)	Format	.sav Variable Name
Machine (1=A, 2=B, 3=C)	1	f1.0	machine
Output	2-4	f3.0	output

CANCER.DAT Exposure to Radioactive Materials and Cancer Mortality Rate (n=9) (Table 5.3)

Variable	Column(s)	Format	.sav Variable Name
Index of Exposure	1-5	f5.2	expose
Cancer Mortality (per 100,000 person years)	6-10	f5.1	mortalit

CLT.DAT 100 Random Samples of Size 50 from Uniform Distribution (Generated by SPSS)

Variable	Record	Column(s)	Format	.sav Variable Name
Marker	1	1-2	f2.0	marker
Sample 1	1	3-4	f2.0	u1
Sample 2	1	5-6	f2.0	u2
.				
.				
.				
Sample 39	1	79-80	f2.0	u39
Sample 40	2	1-2	f2.0	u40
.				
.				
.				
Sample 79	2	79-80	f2.0	u79
Sample 80	3	1-2	f2.0	u80
.				
.				
.				
Sample 100	3	41-42	f2.0	u100

COMPUTER.DAT Gender and Role Portrayal in Computer Magazines (n=661) (Table 12.17)

Variable	Column(s)	Format	.sav Variable Name
Gender (1=M, 2=F)	1	f1.0	gender
Role	3	f1.0	role
1=Seller			
2=Manager			
3=Clerical			
4=Computer Expert			
5=Other			

CONFORM.DAT Husbands and Wives Conformity Ratings (n=20) (Table 5.10)

Variable	Column(s)	Format	.sav Variable Name
Husband's Score	1-2	f2.0	husband
Wife's Score	3-4	f2.0	wife

DEATH.DAT Data on Number of Months Before, During, or After Birthmonth that Death Occurred (n=348) (Table 14.1)

Variable	Column(s)	Format	.sav Variable Name
Month of Death	1-2	f2.0	month

DELINQ.DAT Data on SES, Population Density, and Delinquency for 75 Community Areas of Chicago (Table 6.44)

Variable	Column(s)	Format	.sav Variable Name
Socioeconomic Status (SES)	1	f1.0	ses
1=Low, 2=High			
Population Density	3	f1.0	pop_dens
1=Low, 2=High			
Delinquency	5	f1.0	delinq
1=Low, 2=High			

DIETER.DAT Weights of 25 Dieters (Table 2.10)

Variable	Column(s)	Format	.sav Variable Name
Weight	1-3	f3.0	weight

ENROLL.DAT Data on School Districts, Including the Racial Disproportion in Classes for Emotionally Disturbed Children (n=26) (U.S. Department of Education, Office for Civil Rights)

Variable	Column(s)	Format	.sav Variable Name
District Enrollment	1-5	f5.0	enroll
Percentage of Students Who are African-American	6-10	f5.2	pct_aa
Percentage of Students Who Pay Full-Price for Lunches	12-16	f5.2	pct_lnch
Racial Disproportion in Classes for Emotionally Disturbed*	18-22	f5.2	rac_disp

*Positive index indicates that proportion of African-American students is greater than the proportion of white students.

EXERCISE.DAT Data for 202 Individuals on Exercise Behavior and Health Status (Table 14.20)

Variable	Column(s)	Format	.sav Variable Name
Exercise Category 1=Exerciser 2=Non-exerciser	1-8	f8.2	exercise
Health Status 1=Good 2=Poor	9-16	f8.2	health

FINAL.DAT Final Grade in Statistics for 68 Students (Table 12.4)

Variable	Column(s)	Format	.sav Variable Name
Class 0=Undergraduate 1=Graduate	1	f1.0	class
Primary Language 0=Other 1=English	3	f1.0	language
Final Statistics Grade	5-7	f3.0	grade

FIRE.DAT Data for 28 Firefighter Applicants (Table 2.22)

Variable	Column(s)	Format	.sav Variable Name
Candidate Number	1-4	f4.0	candnum
Sex (1=M, 2=F)	5	f1.0	sex
Race (1=White, 2=Minority)	6	f1.0	race
Stair Climb Time	7-10	f4.1	stair
Body Drag Time	11-14	f4.1	body
Obstacle Course Time	15-19	f4.1	obstacle
Agility Score	20-24	f5.2	agility
Written Score	25-26	f2.0	written
Composite Score	27-31	f5.2	composit

FOOTBALL.DAT Weights and Heights of 56 Stanford Football Players (Table 2.15)

Variable	Column(s)	Format	.sav Variable Name
Weight	1-3	f3.0	weight
Height	4-6	f3.0	height

GAS.DAT Federal and State Gasoline Tax Rates for the 50 States (Table 2.18)

Variable	Column(s)	Format	.sav Variable Name
Tax Rate (cents per gallon)	1-5	f5.2	gastax

HEAD.DAT Head Measurements for 25 Infants (Table 4.7)

Variable	Column(s)	Format	.sav Variable Name
Length (millimeters)	1-3	f3.0	length
Breadth (millimeters)	4-6	f3.0	breadth

INCOME.DAT Income of 64 Families (Table 3.6)

Variable	Column(s)	Format	.sav Variable Name
Income	1-5	f5.0	income

IQ.DAT IQ Scores for 23 Children (Figure 5.1)

Variable	Column(s)	Format	.sav Variable Name
Language IQ	1-3	f3.0	lang
Nonlanguage IQ	4-6	f3.0	nonlang

IQ2.DAT IQ Scores for 24 Children (Figure 5.1 plus hypothetical score)

Variable	Column(s)	Format	.sav Variable Name
Language IQ	1-3	f3.0	lang
Nonlanguage IQ	4-6	f3.0	nonlang

JOBPROF.DAT Job Proficiencies Based on Educational Attainment for 22 Individuals (Table 16.16)

Variable	Column(s)	Format	.sav Variable Name

Educational Attainment
    1=Noncompleter
    2=High School
    3=College
	1	f1.0	educ

Job Proficiency	2-4	f3.0	job_prof

KIDS.DAT Number of Children in 20 Households (Table 2.6)

Variable	Column(s)	Format	.sav Variable Name
Number of Children	1	f1.0	num_chld

LIBRARY.DAT Size of Book Collection and Number of Staff for 22 College Libraries (McGrath, 1986. See Exercise 14.3 of textbook)

Variable	Column(s)	Format	.sav Variable Name
Volumes (100,000's)	1-4	f4.1	volumes
Staff	6-8	f3.0	staff

LUNCH.DAT Data on Calories Consumed at Lunch and Restaurant Temperature (n=14) (Hypothetical)

Variable	Column(s)	Format	.sav Variable Name
Temperature	1-2	f2.0	temp
Calories	4-6	f2.0	calories

MEAL.DAT Data on Restaurant Type, Cost, and Meal Type (n=46) (Hypothetical)

Variable	Column(s)	Format	.sav Variable Name
Meal	1	f1.0	meal
1=Pasta			
2=French			
3=Seafood			
Cost	3	f1.0	cost
1=Inexpensive			
2=Moderate			
3=Expensive			
Chain	5	f1.0	chain
1=Yes			
2=No			

MOVIES.DAT Genre and Gross for 100 Top Movies in 1995 (*Daily Variety*)

Variable	Column(s)	Format	.sav Variable Name
Genre	1	f1.0	genre
1=Action-Adventure			
2=Drama			
3=Family			
4=Comedy			
Gross (Millions)	3-7	f5.1	gross

NOISE.DAT Average Highway Speed and Noise Level for 30 Sections of Highway (Table 5.11)

Variable	Column(s)	Format	.sav Variable Name
Average Speed (mph)	1-4	f4.1	speed
Noise Level	5-8	f3.2	noise

OCCUP.DAT Occupations of 20 Primary Householders (Table 2.1)

Variable	Column(s)	Format	.sav Variable Name

Occupation
  1 = Professional
  2 = Sales
  3 = Clerical
  4 = Laborer
	1	f1.0	occup

PHONE.DAT Monthly Telephone Calls Received by Airlines (Table 4.8)

Variable	Column(s)	Format	.sav Variable Name

Year
  1=1985
  2=1986
  3=1987
	1	f1.0	year

Number of Calls (nearest 100)	2-6	f5.2	calls

READING.DAT Reading Scores of 30 Students Before and After Second Grade (Table 10.2)

Variable	Column(s)	Format	.sav Variable Name
Score Before Second Grade	1-3	f3.1	before
Score After Second Grade	5-7	f3.1	after

RULERS.DAT Age of Death of 42 English Rulers (Table 2.17)

Variable	Column(s)	Format	.sav Variable Name
Age	1-2	f2.0	age

SALARY.DAT Salaries by Gender for 33 Half-Time Clerical Workers (Table 12.8)

Variable	Column(s)	Format	.sav Variable Name
Gender	1	f1.0	gender
Salary (100's of Dollars)	3-5	f3.0	salary

SEMESTER.DAT Data on 24 Students in a Statistics Course (Table 2.21)

Variable	Column(s)	Format	.sav Variable Name
Major	1	f1.0	major
1 = Electrical Engineering			
2 = Chemical Engineering			
3 = Statistics			
4 = Psychology			
5 = Public Administration			
6 = Architecture			
7 = Industrial Engineering			
8 = Materials Science			
Semesters of Statistics Courses	2	f1.0	semester
Grade	3-5	f3.0	grade

SPIT.DAT Data on Success of Interventions to Curb Chewing Spitting Tobacco (n=54) (Greene, Walsh, & Mosouredis, 1994. See page 77)

Variable	Column(s)	Format	.sav Variable Name
Type of Intervention	1	f1.0	interven
1 = Minimum			
2 = Extended			
Outcome	3	f1.0	outcome
1 = Subject Quit Entirely			
2 = Subject Tried Unsuccessfully to Quit			
3 = Subject Failed to Try to Quit			

VOTE.DAT Data on Voting Patterns for 46 Individuals (Hypothetical)

Variable	Column(s)	Format	.sav Variable Name
Plan to Vote in Current Election	1	f1.0	voting
1 = Yes, 2 = No			
Registered Voter	3	f1.0	register
1 = Yes, 2 = No			
Voted in Last Election	5	f1.0	voted
1 = Yes, 2 = No			

WAR.DAT Expectations of Possibility of War (n=597) (Table 11.19)

Variable	Column(s)	Format	.sav Variable Name
June Response	1	f1.0	june
0=Does Not Expect War			
1=Expects War			
October Response	3	f1.0	october
0=Does Not Expect War			
1=Expects War			

WEATHER.DAT* Average Precipitation and Temperature on July 2nd for 78 U.S. Cities. (Data obtained from Internet)

Variable	Column(s)	Format	.sav Variable Name
Temperature	1-4	f4.1	temp
Precipitation	6-9	f4.1	precip

*Note: File "weather.sav" also delineates city name in the variable "city."

WORDS.DAT Number of Words 18 Children Memorized Based on Three Different Experimental Conditions (Table 16.1)

Variable	Column(s)	Format	.sav Variable Name
Information Set	1	f1.0	info_set
1=No Information			
2="3 Categories"			
3="6 Categories"			
Number of Words Memorized	2-3	f1.0	words